Heavy Pack, Rough Terrain:

Letters Home from Vietnam

by

Thomas Houston

Introduction and Afterward

by

Lynn Houston

Formal Feeling Publishing

Copyright © 2025 by Thomas Houston and Lynn Houston

All rights reserved.

This book or any portion thereof may not be reproduced or used in any manner whatsoever without the express written permission of the publisher except for the use of brief quotations in a book review.

Library of Congress Control Number: 2024925539

ISBN 978-1-944355-55-5

For my family

Remembering those who have gone before:

Paul Merton Houston
Dorotha (Dot) Houston
Donna Moseley (née Houston)

Introduction

by Lynn Houston

In the fall of 2016, my father surprised me with this packet of his letters from Vietnam. His gesture came shortly after my return from a reflective summer in Nebraska, spent under the flight path of a Chinook helicopter. There, amidst the echoes of rotor blades, I wrestled with writing about the Vietnam War—a conflict whose stress fed me in my mother's womb and cast a long shadow over my childhood.

As I poured over his letters, I got to the one announcing the news of my conception. The words struck a deep chord, revealing that, yes, I had always—from the first minute—been loved, and that I had been part of a larger, heartbreaking story that connected me to so many others.

What strikes me the most about the voice of the letters is the tenderness or innocence that my father struggles to hold onto over the course of his deployment. He grew up over there. He defiantly planned the rest of his life in the face of death and uncertainty, and then he lived his life just as he had planned it. I think that preparing for what came next must have given him some feeling of control in an environment where survival is often a matter of luck. While there is no Hollywood-style recounting of any major events, these letters are so valuable for what they tell us about the inner thoughts of a homesick young man who is trying to make the best of his situation. In one letter, he talks about "how heavy the pack" and "how rough the terrain," a phrase that seemed to capture not just the physical burdens he endured, but the emotional weight of war—so much so that it became the title of this book.

When my father reads these letters now, he says that he is struck by what a different place the world was then and how totally different his time in the Army was from the rest of his civilian life. He says that he doesn't recognize himself in the letters, and that he gets emotional when reading about the friends he made there. What he finds most surprising about the letters are all the details of the work that he did. He remembered being "just a grunt," but the letters reflect the complexity of his responsibilities. He still treasures the plaque he received for achieving the highest score out of all the recruits in Advanced Infantry Training (which he talks about in the letter dated November 5, 1970). He feels like these letters represent a unique perspective by showing two very different sides of the war: he starts out in infantry and then transitions to rear support after a division stand-down.

He thinks these letters were important to my grandparents because they must have given them some sense of relief, allowing them to know a little about where he was and what he was doing. The news at the time showed very disturbing images of the war. As he was preparing to return home, new recruits coming in from Fort Lewis, Washington, told him to wear civilian clothes on his commercial flight home because there were protestors at the airport throwing eggs and tomatoes at the returning soldiers.

In 2019, for the 50-year commemoration of the Vietnam War, my father was honored with a lapel pin presented by the Garrison Commander of Aberdeen Proving Ground, where I was working at the time. "Welcome Home," they told him, and it felt like the three of us finally were.

August 2, 1970

Hi Mother and Dad,

Well, my first week of basic training is over. I am adjusted to Army living, and I hope the following weeks will go fast.

I am sorry I haven't written sooner, but I have been extremely busy. Sylvia hasn't had many letters either. It was good talking to you today. I wasn't planning on calling you; it was just a spur of the moment thing.

Each morning we have our barracks inspected. We have 57 guys in my platoon, and we are on the top floor. There are four platoons in Charlie Company. Charlie Company is the toughest company at Fort Dix. The barracks are new and made of red brick. The platoon that comes in last for the daily inspection has all the special duties for that day. We came in last one day, and most of us did not get to bed until 11:00 PM. We get up around 3:30 or 4:00 AM and clean. The beds have to be tight and shoes shined and floors waxed. I have seven men in my room counting myself.

This week we were issued our rifle, an M-16. It's very light and seems like a toy; it weighs 7 pounds. We had a class on how to take the rifle apart and clean it. Some of the other classes we have had are Army Drill (how to march), Clergy Orientation, Military Justice, Code of Conduct, First Aid, History of the Army, Field Sanitation, Subversive Groups

Against the Army, and Services for the Soldiers.

This week coming up, we will march to the rifle ranges which are 5 miles away. We will be camping out, also.

Most of the guys in my platoon are really nice. I am just about the oldest. We work together, and so far we are the best platoon. When passes come, the best platoon for that week can get most of the passes. The guys are looking forward to a pass, including me. The passes are only from Saturday 1:00 PM to Sunday 6:00 PM. I will call Friday or Saturday night.

Love,
Tom

November 1, 1970

Dear Mother and Dad,

Well, it is another week down and just two weeks and five days to go.

We had bivouac last week and it was quite an experience. It rained every day we were out. We also had an 8-mile march with a full pack. We had to carry everything we took. We camped at two sites and everything we had got soaked. Pat Manderville was my tent buddy. The "C" rations were not too bad, but like everything, we got tired of them. Breakfast was the only warm meal we had. All in all, it was a good experience, and we did have fun. You are right, Mother, you do have to laugh and joke even in the worst situations.

This weekend we hitchhiked to Baton Rouge. We enjoyed doing this because you meet some interesting people. There were five of us that went. We broke up into two groups. Pat and I in one group, and the other three in another group. It is easier this way to get rides. Then we met at a motel in Baton Rouge. We made a game out of it to see who gets there first.

Pat and I met a very nice young man who is an auctioneer. He lives in Baton Rouge, so we had a long ride. He even bought us coffee on the way. He was a good talker, and we hit almost every subject. Anyway, the other guys beat us to Baton Rouge, but Sunday afternoon we beat them back to the barracks.

It is good to get off post for the weekend. We ate pretty good because of our week of "C" rations. I had veal cutlet and steak.

My plane arrangements are made for November 20. I just hope I can get to the airport on time. I am supposed to get into Elmira airport around 7:00 PM. Do not worry about picking me up. Sylvia may pick me up; if not, Pat's parents will be there. Probably Sylvia will not get into Waverly early enough, since she has school until 4:00 PM. Maybe you may end up getting me. I will have to call from the New York City airport and tell you what time I will be in. We can work something out.

I bought another Reader's Digest.

Take care,

Love,
Tom

November 5, 1970

Dear Mother and Dad,

I appreciate your letters and the clippings of Waverly's football team. Pat also read them, and he knows most of the guys on the team. It looks as if they had a good season. It is about time Waverly's team did something.

What are we doing about Thanksgiving? I think Sylvia and I would rather spend Thanksgiving with you. Maybe John will be there, too. Since I will not be home that long, I would like to spend Thanksgiving with my family.

This week has been a good week. We took our last physical training test. I did pretty good. I got about 450 points. Today we took our knowledge test. We went to nine stations and demonstrated our ability to operate different weapons. I am sure I did a good job. It was rather easy.

The weather here has been changing drastically. In the morning, it is in the 30s, and in the afternoon, it gets around 70. The weekends have been just beautiful. Lately, I have been getting a slight burn on my face.

All of our training is over now. Next week, we spend 7 days in the woods. They call it Tiger Ridge. We go on patrols and play war games. The following week, we clean up the area and process out.

I plan on getting home November 20. I may not be able to, but I will try real hard. If it is at all possible, I would like Sylvia to pick me up. I am sure you can appreciate why. I will have to call Friday night November 20 when I get into New York City and give you the time I will arrive in Elmira.

Everyone is quite excited about getting home. It has been a while since we have seen home.

Donna sent me the picture of fall in Waverly. It is a good picture, and Pat appreciated it also.

Take care and God bless you.

Love,
Tom

P.S. I got another Reader's Digest. It is good reading.

I got this little quote from one of the stories:

> Fear knocked at the door.
> Faith answered.
> No one was there.

December 13, 1970

Dear Parents,

It is hard to get back to Army life after a few weeks of civilian living, but I will make it.

We had a good flight to Chicago, but our flight out of Chicago was one and a half hours late. The mountains are so beautiful in Seattle, especially Mount Rainier. The mountains were snowcapped, but the weather was good. We got a bus into Fort Lewis and arrived about 4:30. We had an orientation meeting at 5:00, and then we got our jungle fatigues.

Today is Sunday, and we didn't do much. We are waiting for our flight to Vietnam. We should leave in about three days. We will have detail and KP until we leave.

I went to the movies this afternoon. It was really good, about college life and demonstrations.

All of my buddies are in the same barracks, so we should ship together.

No address yet!

Take care.

Love,
Tom

Postcard captioned: Buddhist Statues at Hue

Hi Folks,

I got into Cam Ranh Bay 5:30 Sunday morning. We stopped at Alaska and Japan. It took us 19 hours to fly here. So far, I have no return address. I will be here for two days, and then I will be assigned to an outfit.

Love,
Tom

December 21, 1970

Hi Mother and Dad,

Things have happened so fast, I do not know where to begin.

Sunday, when we got into Cam Rahn Bay, we were processed in. We had to change our American money into military issued money. It looks like Monopoly money. All coin money is in the form of paper money. We spent the rest of the day just getting used to the heat and looking around. The mountains were beautiful, and the water in the bay was very tempting.

We went to bed early to get some sleep, but they woke us up at one in the morning and said we were being shipped out. It was an hour flight from Cam Rahn Bay to where I am now and where I will be for another week, Bien Hoa. It is just outside of Saigon. Most of my buddies came with me, Pat and Ed Hamilton. Bob Stevens is not with us; we think he went to the Highlands.

I will not get my permanent return address until I leave here. We will be training here for about one week.

Please excuse the handwriting, I am writing this letter while I am standing in line to be processed into Bien Hoa. I am in the 1st Cavalry Division. I am not sure if that is good or bad; time will tell.

The weather here is pleasant, but I think it will get hotter.

I have adjusted from being a civilian and have accepted the fact that I am in Vietnam and I will do my best. Even though it still doesn't seem possible.

We all have good spirits, and I'm sure everything will be okay.

The drinking water here is terrible, so we drink Coke or beer.

The barracks are made of wood, and even though they are not anything to look at, at least they get you out of the weather and you have a mattress under you.

Today I took my first malaria pill, because Monday is the day of the pill.

Take care and God bless you.

Love,
Tom

P.S. Do not write me using this return address. I will send my return address as soon as I get it.

December 25, 1970

Dear Mother and Dad,

Today was a different Christmas than the ones I am used to. Thank you for the gum and pocketknife. I shared the gum with my buddies.

We did not have training today. It was nice of them to give us a day off, but there was little to do. I did play some basketball and volleyball. In the afternoon, they had a rock band, and they were very good.

It is not every Christmas that you can lay out in the sun and get a tan. The sun gets quite hot, so I will have to be careful. Even though they had Christmas trees and decorated the mess hall, it did not seem like Christmas.

Yesterday when we were in training, we had a church service. It was pretty good; we sang Christmas carols and had a little sermon.

Bob Hope was just a few miles from here today, but we could not leave our post.

I hope in three or four days I will be able to send you my address.

The training post that I am in now is quiet. We are near a helicopter base, so it does have a little traffic overhead.

We had a really good Christmas dinner. Turkey was the main course and all the other things that go with turkey.

I hope Donna and Charles enjoyed their visit. I imagine it was good to have them home.

Christmas Eve everyone did his own thing. I drank soda and sat around talking with my buddies.

Right now, when I am writing this letter, it is 6:00 AM Christmas morning at home. I can't help but think about all of you on Christmas. I hope you enjoyed your Christmas, because I did considering the circumstances.

Take care and God bless you.

Love,
Tom

P.S. I made an allotment out to you, Mother, to cover the $25 a month for rotary. It should start the end of February.

December 27, 1970

Dear Mother and Dad,

Yesterday I spent all day at the Bien Hoa airport waiting for our flight to Soung Be, pronounced "Song Bay." It is up by the Cambodian border. This is my permanent assignment. Pat and Ed are up here with me, but they are in a different company.

My address is the following:

>PFC Thomas P. Houston
>084-36-4125
>Co. D 5th BN 7th Cav.
>1st Cav. Div. (AM)
>APO San Francisco 96490

The (AM) stands for Airmobile. This means we are transported everywhere by helicopters.

Today is my first day here, and I am sitting in a tower guarding the fire base. There is very little action here. I hope it stays that way.

Please forgive the dirty stationary. There is red dirt all over this place. This firebase isn't much, but it will be my home for a while.

While I was at the first team academy, we learned about helicopters. The last day there we had to jump off a 50-foot tower backwards. They call this rappelling. It has to do with a rope seat and sliding down the rope. I was not too excited about it, but it was easier than it looked.

The guys up here are very friendly, and they help us new guys out considerably.

The time will go by fast. We spend about 15 to 20 days in the field, then we come back to the fire base for a few days, then back out we go.

January 1, I will be going back to Bien Hoa. My company has been out in the field, and they get a few days to stand down. This means they can enjoy a few more comforts of life.

I hope you all had a wonderful Christmas and happy New Year. For me it doesn't seem like Christmas or New Year, but I will celebrate it when I get home.

I really feel sorry for these people. They have nothing, except flies and mosquitoes. I hope this war ends soon.

That's it for now, will keep you posted.

Love,
Tom

January 2, 1971

Dear Mother and Dad,

I hope everything is okay with you. It has been a long time since I heard from you, and I can't help but wonder how everyone is back home.

I am still at fire base Soung Be. We new guys didn't get to go to Bien Hoa. It is not a bad life here. We pull perimeter guard at night and sleep during the day. The only problem I have is trying to sleep during the day; I am not used to it.

Since we have most of the day free, I have been doing a little reading. I read a book about consumers, and I just started one on the stock market. I am trying to keep my mind active.

You would not believe the change in temperature here. During the day, you sweat, and at night, it gets cold. Every night on guard I have worn my field jacket.

A lot of the details are taken over by the Vietnamese. This is good, because I do not like KP.

Pat and Ed have gone to another firebase called Snuffy. When my company gets back from Bien Hoa, I will be going there also.

If the Army should send my W-2 form home, please forward it to Sylvia. I told her if she had trouble filling out the return, you would help her, okay Dad?

Hey Dad, could I borrow your hunting knife? They did not give me any, and I would like to carry one. I think it is in the back room in one of the drawers.

Mother, please do not send any type of clothes. The Army will give me what I need. Besides, I have to carry everything I own. They will even give me clothes to come home in.

Could you please send me John's new address? He is the only one I have not written to yet.

I am fine and have started counting the days. This place is not the most luxurious place I have been in.

Love,
Tom

January 2, 1971

Dear Mother and Dad,

I got your letter yesterday (#9) and also one from Donna and Marv. I was going to answer your letter yesterday, but I had a headache and laid down. I guess I got too much sun. I feel better today and wanted to mail this letter before we go to the field.

We were at Sundy Punch for nine days. We pulled perimeter guard and other little details. Today we left by troop carrier (helicopter) to firebase Audey. We probably won't leave here until 1:00 PM. We leave by helicopter. These are smaller ones; they carry about four or five of us. Since I am the Machine gunner, and I have the most firepower, I am going on the first bird with the captain. My buddy who I met at Fort Polk and came to Vietnam with is my assistant gunner.

You asked about my duties. When I first got here, I was a rifleman. This involved carrying an M-16 rifle. Every night when we set up our perimeter, I would set up my trip flair on my Claymore mine. Now that I am carrying the M-60 machine gun, I no longer carry a trip flair or Claymore. Since I have the most firepower, I walk about 3rd man back from the point man. The point man clears the path and walks on a set azimuth. Other than this, I have no special duties. We all have to pull guard about one hour each night.

I think we will have about two more small assignments before we will be sent to a different unit. This is when we go to Bien Hoa for Brigade stand down.

I hope my pictures turn out okay. Some of the other fellows have a camera like mine and they have some pretty good shots. Marv and Sharon have moved into a new house. If they like it, they have an option to buy it. Right now they are just renting.

We do not have a problem with drugs in my platoon. We travel in platoons out here. I am in the 3rd Platoon; this consists of 28 men. There are some guys in my platoon that do smoke "grass," but they do not do it in the field. We have a good bunch of fellows and our LT is all right, too. He is 25 years old and takes care of us.

Since this may be our last time out in the field for the Cav, the LT gave us a pep talk. He said we are going to play it cool until we go on stand down.

Not much more to report. I am fine and there is relatively little action where I am.

Since I will be in the bush, my letters will be a while before they get to you. Every three days.

Take care!

Love,
Tom

January 12, 1971

Dear Mother and Dad,

Things have happened quite fast since the last time I wrote you. I left Soung Be by airplane and flew to a small fire base called Snuffy. Here I spent two days getting my weapon zeroed in and getting the rest of my equipment. From there, I flew by helicopter to an even smaller fire base called Audey. Here is where I finally met up with my company. One of my buddies that I met at Fort Polk is with me and we shared the same tent. The next day we took a helicopter to Sundy Punch, and from there we started walking in the jungle.

I have been in the jungle for two days now. I do not believe how heavy the pack is and how rough the terrain we have to go over. We have to carry enough food and water for three days. On the third day, we get logged. This means clean clothes, one hot meal and more water and also our mail.

We carry "C" rations and also heating tablets to heat our meal.

Since it is warm out, we just string our mosquito nets between two trees and sleep under that.

So far, we have not hit any contact. I hope it stays that way.

The guys in my outfit are very friendly and helpful. We have a young LT, and he is okay, too.

We will not get out of the woods for at least another 20 days.

January 20th is Sylvia's birthday. I wrote her a letter and I hope it gets to her in time. We have plenty of time to write here, but very little to say and they only collect our mail every three days.

I will try to keep you posted. It sure will be good to get some mail. I should get some tomorrow. I got one letter from Sylvia already.

Take care,

Love,
Tom

January 15, 1971

Hi Mother and Dad,

Not much to write about. We do the same thing every day.

I got your letter and one from Donna and Sylvia last log day. Log day is a little like Christmas, after being in the woods for three days. We get our one hot meal, clean clothes, more water and more "C" rations. They also gave us shaving equipment and candy. I sure do look forward to log day, so I can get letters from home.

I am growing a moustache just to see what I look like with one.

We are in the Central Highlands and I sure am getting tired of walking up and down mountains with my pack.

So far, we have not seen anything. That is just fine with me.

We have been following a trail to see what is in the area. The trail has not been used for at least four months, so there is little activity in my area.

Tomorrow is log, and that is when I can get my letters out. Not much to write about, but I did want you to know that I am fine.

Besides "C" rations, they gave us LRP's (Long Range Patrol food). It is dry and all you have to do is add a half cup of hot water and it is ready to eat. So far I have had spaghetti and beef stew, and they are not bad.

Thank you for John's address. I got a letter off to him and clued him in on my activities.

It sounds as though you all had a nice Christmas.

Take care and I think of you often.

Love,
Tom

January 17, 1971

Dear Mother and Dad,

Yesterday was log day, and I got two letters from Sylvia and two from you. I cannot begin to tell you how much I appreciate your letters. Just to hear about home and what you are doing is important to me. It was pretty hot Sunday out on the LZ (Landing Zone) yesterday. We have to chop down trees with machetes and clear an area for the helicopter to land. I got a slight burn on my arms and neck, but tonight they are fine. When we are walking through the bush, not much sun gets through, but log day is when we are in the open and it gets quite hot.

To give you a better idea where I am, we left Sundy Punch one week ago today. We have been traveling in a Southwest direction.

From Sundy Punch, we can see into Cambodia, so it is quite close to the border. I think we are heading towards Snuffy, but it is hard to tell. I think we will be in the woods for another 20 days.

I am fine and getting in shape very fast. The only thing that has been giving me trouble are ticks. They get in your skin and start sucking your blood. When it gets really hot over here, they say the leeches are bad. [A dog on base often had leeches on his skin.] I have my buddy check my back twice a day now and spray insect repellant on me. They are not as bad now. I hope I have gotten rid of them for good. I still have some sore spots, but I think they will go away.

The word has been going around that the 1st Cav is going home in March. This means I will either go into another brigade of the Cav or go to another unit. Time will tell and either way, I will let you know.

About my care package! I really do not know what to suggest. I do not have much, so anything would be a blessing. Here are a few suggestions: maybe some small cans of Start or Koolaid, maybe some small cans of meat (ready to eat), cheese in a pressure can and some onion salt to flavor up my LRP's and "C" rations. Actually your guess is as good as mine. Anything to eat would be appreciated. Please do not send me a lot, for I do have to carry it. On log day, we do not move, so I can eat the bulky items and I also share with a few of my buddies, for they share with me. I do have some room in my pack for good food and I am getting a little tired of "C" rations.

Mother and Dad, I really do appreciate all you have done for me. I get a lot of time to think here in Vietnam and as I look back on the years, you have been ideal parents. I have thanked God many times for having you as my parents and now I am thanking you. You have made me strong and gave me a wonderful faith in God, and I can face whatever I have to in the days and future to come. One never really appreciates what he has until he gets to a place like this and believe me I love you both.

Today was just another day. We did wade through two streams. The water got up to about our thighs. We were on the move most of the day. We had a hard time locating a decent spot for our log tomorrow. I think we called in combat engineers to blow a landing spot for the helicopters.

Here are a few things I could use: Dentyne gum, a small tube of First Aid Cream, a small plastic bottle of medicated body powder, and a small tube of Crest toothpaste. I think that is all for the time being. Oh yes, a toothbrush (hard).

That is about all for now. I do enjoy your letters, so write when you can.

Take care, you are in my thoughts.

Love,
Tom

Army band flown over from the States to entertain the troops

January 24, 1971

Dear Mother and Dad,

Yesterday was log and I received your letter dated January 15th. I also got your care package, and thank you very much. The knife is in beautiful condition, and I sure can use it. Dad did a good job on the knife.

Mother, I would appreciate it if you would not send me candy. I get enough sweets over here. They are not good for my teeth, and I have enough trouble brushing them every day. Send me fruit or small cans of ready-to-eat food.

I also got a care package from Sylvia. She sent brownies and oatmeal cookies. They were very good.

We have been in the bush for 15 days. I think tomorrow we will go to a fire base and guard the perimeter for about a week.

I got a nice letter from Uncle Dick and Aunt Polly. They sent a picture of Donna and Charles and John at your get-together at Christmas. I think they really enjoyed John's snowmobile. It was also good that they could have their family together for Christmas.

You know, Mother and Dad, I think this experience in Vietnam is doing me good. It is hard to explain until someone has gone through it. It has brought me closer to people and things and made me appreciate them more. I have been closer to God since I got out in the bush. You look to your buddies and God to look over you, for they are the only ones that can help you. I have started reading the New Testament in a pocket size Bible that was given to me when I was inducted in Syracuse. It puts my mind to rest.

I have heard that some states give money to servicemen that served in Vietnam. Could you check if N.Y. state does and how much they give? Some states give as much as $600.

The chaplain was here yesterday and we had a short service. It has been a while since I saw him last.

I wrote to Pat Cary to see how her family was and to find out if she knew what some of my high school friends were doing. She wrote back a nice letter about her family and about what some of my friends were doing. It was good to hear from her.

The pictures of Gwenda Alfred and her husband were very good. I am glad she was able to come back to Waverly.

That is about it for now. I will be careful.

Take Care!

Love,
Tom

January 27, 1971

Hi Mother and Dad,

Thanks for the letters for they are coming in like a charm. Also, thank you for the Teen Power [a newsletter in pamphlets]; they made for very good reading.

We are still at Sundy Punch and it looks like we will be here for a few more days. A helicopter was bringing supplies from Snuffy to Sundy Punch and dropped it in the jungle accidentally. We have been out twice, but unable to locate it. It is like trying to find a needle in the haystack. They can't locate it by air, so how do they expect us to find it on foot?

Maybe I didn't tell you, but Pat and I are in the same battalion, but I am in Company D and he is in Company C. I got a chance to talk with him for a few minutes about two days ago. As we were leaving the field, his company came in to take over our position. As he landed and I was waiting for my helicopter, we had a few minutes to talk. He said he heard from Pete Abatiell (the one we visited in Vermont). He is in a different battalion of the Cav. I will probably see Pat again in Bien Hoa when we go on a battalion stand down in March.

I bought an Instamatic camera over here. It is an X-15 model and you do not need batteries for the flash. I got a good deal on it. Do you think I should take slides or prints? I think I will send the prints to Sylvia, so she can look at them and then the slides to you so you can show them. You will have to send the film to me, for all I can get here are the prints. Send about two packs each month, okay. It takes Kodachrome-X 126 cartridge (20 exposures). I will then send the film back to you, so you can get it developed. It is easier that way. Sylvia can give you the money to get them developed.

I got a nice letter from Norris Rock and he sent me $20. He sure is good to me. [This was his supervisor when he was the manager of the swimming pool at Bloomsburg.]

I am glad Dad is getting out and seeing the basketball team. It is also good that he has a buddy to go with. I imagine John Howe appreciates it also. He is a good fellow.

They have me carrying a machine gun now. It is not any more dangerous than carrying a rifle, so don't worry. It is just heavier and that is all.

I heard on the grapevine today that we will not be moving much in the field. We will just set up a camp and wait.

Most of the time we talk about what we were doing before we got drafted and what we are going to do when we get out. I have been doing quite a bit of thinking about my future and I think I have it pretty well worked out.

You are right, Mother, I am not out here alone, for I have a lot of friends back home praying for me.

I do not know if I mentioned it or not, but I got a nice letter from Uncle Norm and the family. He told me the story of how he happened to go to Buffalo and work with Ford and how he planned his future then, just like I should be doing over here. [Uncle

Norm had bad asthma; he had to move to Buffalo because of the weather, when all of his family and friends were in Waverly.]

I am fine, and everything is calm. The ticks have left, so I am as good as new.

Take care and I will keep you posted.

Love,
Tom

Log day at base camp, February 1971

February 6, 1971

Hi Mother and Dad,

Happy Birthday, Mother. I am sorry this birthday greeting will get to you late, but I am thinking of you on your birthday, as I do every day.

I got letter #10 and #11 and also a care package with nuts and crackers. They sure were good.

This is my typical day: we are woken up by the last man on guard about 7:00 in the morning. We have to be rolled up, eaten breakfast and ready to go by 8:30 A.M. We generally work for one hour and take a 15 minute break. We do this until around 11:30 a.m. Then we form up a perimeter (circle) and eat lunch.

We moved out again at 1:00 PM and looked for a NDP night defensive perimeter. It has to be on high ground. At 2:00 PM, we normally find one and set up our own perimeter, two men to each spot. Then one squad goes out on a cloverleaf and checks out the area. The other squad stays behind and pulls security guard. The squad that goes out on the cloverleaf goes out "light." This means no rucksack, but they do carry their weapon, ammo and smoke grenades. When they come back, everyone runs out his Claymore mine and trip flare from his position. We are usually set up by 3:30 or 4:00 PM. Then when everyone has their set up, the LT gives the word to clear an area to sleep for the night. Then we usually eat supper and read or write letters until dark which is seven or 7:30 PM then we go to bed and everyone pulls an hour guard. We usually have a guard position for each squad.

We have very good noise and light discipline. At night there are no lights, not even a cigarette is lit. When we walk in the bush, no one says anything unless it is important, and then it is only a whisper. Even when we are set up for the night, we still have to whisper. The only time that I have talked in my normal voice is when we are on a fire base.

There isn't much twilight here. Nothing like back in the states.

My buddy and assistant gunner is from Michigan (Kalamazoo) and his name is Hal Henning. He has two older brothers and two younger sisters (twins) who are seniors in high school. He went to college two years before he got drafted. He will go back and finish up when he gets out of the service.

I am the third oldest in my platoon. The LT is 25 years old; then we have a platoon sergeant who has been over here for two years. I do not know how old he is, around 36, and then me, 23. Most of the guys are 21 or 22.

Dad, your knife is working out great, but the side of the blade that goes against the sheath is getting rusted. I tried everything to get it off, but it did not work. The other side of the blade is fine. Can't figure it out!

I am reading a book about the stock market. It is quite good, and it keeps my mind active.

Dad, could you help Sylvia figure out our income tax? If she needs help and also check in my New York State tax. I wish I was home to settle these finances, but she will have to do the best she can. The Army made a mistake on my W-2 form, so I am going to find out about it.

Today is Sunday (7 February), but it is just like any other day. The chaplain had a service last log day. Take care.

Love,
Tom

Hal Henning, assistant gunner

February 9, 1971

Hi Mother and Dad,

I got letter number 12 yesterday. I hope dad is feeling better. He has to be more careful.

My car registration is due at the end of March, and it also has to be inspected in March. Sylvia will have to come up and take care of these two items. We did get our money back from Pennsylvania when I got my Pennsylvania plates and then gave them back. I sure wish I knew where we are going to settle down, so I could get the car registered in that state and have my license in that state. I also do not like the idea of paying New York State tax. It sure is a big problem to have your wife living in Pennsylvania and me in New York. When I get back to the states, I will still have some time in the service, so I do not know if I will take Sylvia with me or not when I find out where I will be stationed. She has a good job, and I would not like to see her give it up. I guess time will tell.

I got a nice letter from Mrs. Grenel (the lady from church). It was good to hear from her. I also got a letter from Coach McLaughlin [his swimming coach] and Coach Turberville [the head coach when he served as assistant coach]. One of our best swimmers for BSC broke his finger and coach is quite upset. Some of the football players for BSC took a list of grievances to the president of the college about the head football coach. The big brass had a meeting and he was told to resign. Coach Puhl my track coach is now head football coach. They sure are having their trouble in the physical education department at Bloomsburg.

I got a letter from Jim McBride and he is doing really well at Penn State. He likes it much better than Bloomsburg. He hopes to have his Masters by the end of the summer.

Not much new is happening with me. We have a scout dog with us and his name is Teddy. He weighs 80 pounds, but is gentle, although he can get mean. He is good to have around and he is in front when we go down a trail.

After being in the jungle for a while, you are able to improvise. Our mosquito net is in the shape of a pup tent and we use plastic spoons we get from C rations for tent pegs. I also have been trying to doctor up my LRP's. I add cheddar cheese that I get in my C's to my beef stew and also the beef broth you gave me adds a pretty good taste.

We get soda rations on log day, which we pay five dollars a month. They also give us some ice usually.

I actually do not mind the jungle, if it wasn't for the danger involved. I enjoy sleeping in the open and more or less camping out. This is most definitely roughing it. That's about all I have. I am fine and well. I have good spirits, and I'm getting in shape.

Take care!

Love,
Tom

February 11, 1971

Hi Mother and Dad,

Yesterday I got your 13th letter. I also got a Valentine's card and a little note from Aunt Ferne.

I hope you had a nice birthday, mother!

Not much has happened to me. We stayed at the same camp today as we used yesterday. My squad went on a patrol today, but as usual we did not see anything. We spent about an hour walking down a stream and right now my boots and socks are drying up. The rest of the day has been spent just relaxing and reading and writing letters.

I figure in about a week we will be going to Bien Hoa for brigade stand down and from there we will be assigned to our next unit.

On log day, we got a newspaper that the Army puts out. It is called the *Stars and Stripes*. It has news from the world and about the war. It looks as though there will be an offensive in Laos, but so far it has not affected me.

One of the guys from my squad was taken to the hospital in the rear yesterday because a bug bit him and hurt him pretty bad. The medic took care of him, and I think he will stay with us. For me, I am fine and well.

I am not too sure what I meant either about states giving men money who served in Vietnam. I heard that they gave them money out right, but I guess that is pretty improbable.

I will write tomorrow.

Today is the 12th, and we did very little. We moved about 600 meters from our last campsite. Tomorrow is log, so we will use the same log pad as last time.

I think the hot season is coming fast. It is getting hotter during the day, and it is not as cool at night. The sweat just pours out of me now.

I am reading a book called *Fathers and Sons* by Ivan Turgenev. It is one of his best works. I got plenty of time to read over here and I think I will read as much as I can. Once we get settled at night, that is when I write my letters. I usually get a chance to answer my letters.

This is about all I have for now.

Take care!

Love,
Tom

February 14, 1971

Hi Mother and Dad,

I got your Valentine's card and letter number 14. Thank you very much. I also got a card from Aunt Polly and Uncle Dick, Donna, and Sylvia.

Things happen very fast out here. Right now I am on fire base Audey. Last log day we were waiting for log on the pad and the CO said we were going in so the birds came and got us. I think we will stay here for about a week and then go back out in the field for a little while, but one cannot be sure. This firebase is going to be torn down in about three weeks, so we are bringing the perimeter back further and building temporary bunkers. [He was the last man to leave because he had the bigger machine gun.] It is extremely hard on the fire base and there is no shade at all. At least when we are in the bush we have plenty of shade.

Actually, I cannot use American money over here. The military have their own type of money that they use. I can get it converted when I am in the rear, but I think I will use it to come home on my leave.

I got my camera by chance. When I was on Sundy Punch, they had a PX. They happened to have a camera, so I got it. I really cannot get too many action shots, because I need both hands on the machine gun. I am getting used to carrying it and it is not so bad now. About my future! Sylvia and I have been making plans, but a lot depends on how things work out when I get out of the service. I got a long letter from Coach Houk [Athletic Director of

Bloomsburg University] and he said he would like me to be assistant track and swimming coach. I would like it, but first I want to get my Masters. Sylvia also wants to get hers. This is a goal we have set. Coach Houk has had a lot of injuries on his wrestling team and they are not doing so well. If I do not get a job at Bloomsburg, I will try to get a teaching job at a junior college either near Bloomsburg or around the Waverly area, Corning. I will probably go full-time to finish up my Masters when I get out of the service. Right now we are saving for our home. The future is still uncertain, but we have set our goals. A lot depends on where I can get a good job.

I got a letter from Jim McBride, and he is doing fine. Sandy and Jim plan on getting married this summer.

It sounds as though Donna and Charles have solved their problem and things are working out for them.

It is getting dark, so I will end for now. I will keep you posted.

Take care!

Love,
Tom

February 15, 1971

Dear Mother and Dad,

I got letters number 15 and 16 and a care package today. I also got cookies, Gatorade, and peaches from Sylvia. Thank you for the things I asked for; they were what I needed.

I guess I didn't explain it too well to Sylvia, but I want to file income tax jointly for 1970. I want to take advantage of the income splitting tax advantage because we are married. Help her if she needs it. Also, she said that there is a noticeable rust spot on the rear fender [of her yellow Pontiac Firebird]. Could you look at that when she brings the car up?

I also got letters from Uncle Norm and Aunt Joe, Henry and Grace, and Mr. Ross from college [Bloomsburg University Economics professor]. The swimming team beat Glassboro, but lost a close one to East Stroudsburg, 70 to 65. Dave Gibas, the outstanding sprint swimmer, broke his finger but he is swimming with a splint on it. Sorry to hear about John [motorcycle accident], but I guess it could have been worse.

Not much is new with me. We took down barbed wire today. [One of his buddies was trying to pull it loose and got cut.] At least they gave us heavy gloves to wear. We also built a shower and it sure felt good to relax and get clean. Last night, I saw a Bonny and Clyde type movie. It had lots of action and I enjoyed it. My buddy Hal stayed and talked while I was on guard, so it went pretty fast. We talked about everything.

It was rather cloudy today, but it was still warm and pleasant.

It sure sounds as though you are having a hard winter. I sure wish I could trade you for a while. Sylvia has been doing a good job in the savings department, and I am proud of her. I guess it is easier to save when you have a goal.

That is about all I have for now. Take care and stay healthy.

Love,
Tom

February 20, 1971

Hi Mother and Dad,

I received letter number 17. I also got one from Reverend B., Mr. Barry [a student teacher] and Mrs. Allmen [his former landlord in Bloomsburg].

We stayed at Audey until the 19th. Yesterday, the whole company was flown out by helicopter to a spot in the bush. From there we split up into our platoons and went different directions. After spending a week on the fire base, it sure was rough walking in the bush with a heavy pack. My old track injury came back. I had heel bruises on both heels [because they gave him boots that were too big]. It made it difficult to walk. The medic looked at it yesterday and we decided that I should go back to the fire base. The helicopter picked me up this morning and I am on Snuffy. [LT was not happy about the noise the chopper made, and he missed his mail because it was coming in while he was going out.] The doctor looked at them and said I should stay off my feet as much as possible. Other than my heels hurting, I am fine. I imagine they will give me some small detail while I am here, but other than that I can read and write letters. I will stay here until my platoon comes in for stand down, which is in six days. Then we will go to Bien Hoa and get assigned to another unit.

I had a long talk with another one of my buddies from California. We talked about business, because he was attending a junior college and majoring in business administration. He thinks he wants to go into insurance. I am fine other than my heels hurting. Not much more to report.

Take care!

Love,
Tom

February 23, 1971

Hi Mother and Dad,

Well, I am back in the field. My heels are somewhat better, and I got tired of doing details, so I decided to go back to the bush. The guys were glad to see me, and it was good to be back. Actually, I got as much rest in the field as I did on the fire base doing details. I just missed my mail on log, so I will get it tomorrow.

They told us to write to our loved ones and tell them not to write or send any packages until we get our new address from our new unit. It will be the same old thing again. I will be able to write, but you won't.

We are not doing much in the field now, just taking it easy and ready to go to Bien Hoa for stand down.

When you write to Donna and John tell them to stop writing and I will send my new address when I get it.

Not much more to report. I am fine and feel good.

Take care.

Love,
Tom

Cleaning M-60 at forward fire base

March 1, 1971

Hi Mother and Dad,

I have been unable to write lately because they have kept me quite busy and moving from place to place.

On the 26th we were flown out of the field to fire base Audey. We stayed there for one day and tore it down completely. The only thing that remained was a large dirt pile on top of the hill. Then we flew to Snuffy. We stayed at Snuffy for one day, trying to get an airplane to Benoit. While we were waiting for a flight, I visited with Pat and Ed. It sure was good to see them again. We got into the VIP center in Bien Hoa about 7:30 PM. Pat and I played pool and we beat everyone we played. It sure boosted our morale. We also saw part of the movie. I had seen it before. The sleeping conditions are very good compared to the bush. We have bunk beds with mattresses. We got a shower, and they give us clean clothes. The VIP center is a good place to relax. We only have a few details. They also have a swimming pool we can use.

I got my assignment. I will be going to the 101st Air Mobile Unit. Hal and a few of my buddies are going with us. Ed is going to an Americal unit and Pat is staying with the Cav. I don't feel too bad about the assignment, for one unit is as good as another and Vietnam is Vietnam.

When we got on Audey, I got your care package. Thank you for the film and goodies. I also got letters up to number 22.

When the cav moves out, there will only be a brigade of American troops in this area. The South Vietnamese soldiers are taking over the fire bases. Since Audey and Sundy Punch were small ones, they tore them down. They did take over Snuffy.

I question my W-2 form that I got from the Army. I had the clerk check it, but he could not find anything out about it. I am sure they took income tax out for me, but they did not show it. If they did, it would only be a small amount, so go ahead and have Dad file Sylvia and I jointly. I know I do not have to file this year, but I want to get it out of the way, and I want to file jointly with Sylvia, so we can save.

Yes, Sylvia and I would like to have a garden when we settle down. It would be a small one like Dad's, but you sure can save some money on food bills, and it is good to work in the soil and a worthwhile hobby.

When I get my new address, I will send it. Not much more to report. We have a party this afternoon (steak). I have to report to the 101st March 6.

I also got letters from John, Cheryl and George, and the Smiths. [Cheryl Hoer is his cousin, the daughter of Aunt Polly and Uncle Dick Dickinson, and George is her husband; the Smiths refers to his Aunt Joe and Uncle Norm.]

Take care

Love,
Tom

March 4, 1971

Dear Mother and Dad,

I am still at the VIP center in Benoit. Tomorrow I start processing out. I am supposed to leave for the 101st unit on March 6, but because of the Laos situation, it is hard to get flights out. They have been sending the guys to the DEROS center, and they wait there until they can get a flight out.

It is pretty nice here, and you don't even know there's a war going on. They have places to play basketball, softball, and a swimming pool. We are free most of the day, so I have been swimming a lot.

Pat, Hal, and I usually get together and do something. At night, they have a live band and usually two shows.

I called Sylvia yesterday. It cost about $20. It was good to talk with her. Because of telephone security, I could not tell her where I am or where I am going. It takes a while to call, and actually I do not think it is worth it, so I doubt if I will do it again.

Pat said he called his mother the other day.

I had two cheeseburgers yesterday, and they sure tasted good.

I think it will be a while before I get my new address, but I will send it as soon as I get it.

I got your care package the other day; fruit, nuts, and Dentyne. They had to repack the package because the box broke open.

I sent Sylvia two catalogs, so she will send you one, and if there's anything you would like, tell me, and I can order it for you. Also, I think John might want some stereo equipment.

I really don't know where I will be going, just that I am assigned to the 101st unit.

Pat is staying in the Cav and Ed is going to the Americal.

I got a nice letter from Pete Abateille. He will remain in the Cav, because his brigade is the one that is staying here.

Not much more to report. I am fine.

Take care,

Love,
Tom

March 8, 1971

Dear Mother and Dad,

I will try to tell you what I've been doing since the last time I wrote.

The 6th of March, I processed out of the Cav, and then from there they took us to the Bien Hoa airport. We finally got our flight out at 3:00 PM. I saw Pat at the airport and we said goodbye. From there, he is going to Phuoc Vihn. The plane we were on had engine trouble, so we had to go to Saigon. They got it fixed, and we arrived at Phu Bai at about 11:00 PM. If you check on your map, Phu Bai is as far north as you can go. We are in the process of getting our records in order. I am now a proud member of the 101st Airborne Division, otherwise known as the screaming eagles or the bald buzzard. I met a friend of mine here that I met at Fort Polk: Paul Cable. He is a clerk typist and he is trying to get me a rear job clerk typist. I would appreciate anything he can do for me. I should have my interview tomorrow or the next day. Hal, my buddy that was in the Cav, is trying to get into the MPs and he has a good chance.

Phu Bai is a pretty nice area. It is in the rear and very little action. The forward fire bases are further up north.

It looks as though I will have to make some new friends. Only a few of my buddies came up here from the Cav. Oh well, this way I meet different people and learn from their experiences.

I sent a 1st Cav Division yearbook home to you. I thought you would enjoy it. It has some good pictures and gives a history of the 1st Cav Division in Vietnam.

I hope you are getting the $25 that I have sent home to you for my rotary loan. It should have started at the end of February.

The money that I make over here I decided to send it home so Sylvia can put it in the bank. I will only draw the money I need to last me the month.

By the way, I earned my CIB combat infantry badge and Vietnam service ribbon and medal. Not much more to report. I will let you know how I made out on the rear job.

I still do not have my permanent address, so please don't write, for I won't get it anyway.

Take care.

Love,
Tom

March 10, 1971

Dear Mother and Dad,

Well, I got my rear job. I am working as a clerk in the casualty branch of the 101st Administration Company. We are stationed at Phu Bai where all the records are kept for the guys in the field.

Today was my first day on the job, and it sure is different from the field. We work every day from 7:30 to 5:00.

This is how I got the job. When I went in to get my records processed, I asked if I could take the typing test; they let me, and I did real well. The captain of the casualty branch called me in and saw that I had a B.S. in Accounting and that I am tall to help them out in their volleyball league, so I got the job. I live in a barracks with six other guys. I have my own bed and two wall lockers. We even have a small refrigerator to keep things cool. It is nice here.

I played volleyball for them tonight and we won, so we are now 5-0. They also have a softball game coming up.

Hal, my buddy from Michigan, is trying to get into the MPs, but I do not know how he made out. This is my address:

PFC Tom Houston
084-36-4125
101ST admin co-casualty branch
APO San Francisco 96383

The other night they had a live band from Miami, Florida, and they were pretty good. They also have movies every once in a while.

I probably will not be taking many pictures with my camera anymore, since I will be working most of the day.

Thank you for the picture you sent me; it is a good picture of both of you. I think I forgot to mention it when I received it.

Here are a few suggestions for my care package: Premium crackers, peanut butter, orange juice and cheese in a pressure can.

The guys that I work with are nice and very helpful. They are younger than I am, but we get along real well.

Not much more to report it sure it will be good to start getting mail again, for letters mean so much.

Take care.

Love,
Tom

March 12, 1971

Dear Mother and Dad,

Today was a pretty good day. I was busy, so it went fast.

The first few days on my job I was just learning how to take casualty reports and learning the procedure of the office.

Today I moved to the job that I will have while I am here. I will be writing sympathy letters to the wives and parents of their husbands and sons killed in action over here. I really do feel sorry for the people that have to read these letters. There's a lot of paperwork involved and I am typing most of the day. We have had a lot of missing people lately, because of the helicopters shot down over Laos. These letters are hard to write to the family, because we have very little details to go on. I hope this war ends soon, for there is too much killing on both sides.

Tonight we had a practice softball game and we lost. I really enjoyed the game, for I haven't played much ball for a long time. This life sure is different from the field. I can't help but think of the guys in the field, for I know what they are going through.

The guys I work with are very nice and I didn't have any trouble getting to know them.

The food they serve here is really bad. I guess I will have to get something for our refrigerator. I have been buying cheeseburgers every night, but that gets pretty expensive.

After working in the office all day, I have to get out and get some exercise. They have a pretty good sports program.

That is about all for now. Take care and God be with you.

Love,
Tom

March 16, 1971

Hi Mother and Dad,

I am getting the feel of my job better, but there is a lot to learn. I have plenty of work, so the day goes fast.

It sure will be good to get some mail again. I have been writing a lot of letters, to let everyone know my new address.

The guys in my barracks are very nice and we get along good. I didn't have much Army clothes when I came up here, so they have been giving me some of theirs.

I have gone to a few movies since I have been here, but the sound is bad and you can't hear what they say, so I don't stay long.

It is raining today and has been cloudy and cool all week. It is different from down south.

I have bunker guard tonight, which is just one of the many details we have to perform in the rear.

Not much more to report.

I am well and think of you. Take care.

Love,
Tom

Office work at the casualty branch of the 101st

March 21, 1971

Dear Mother and Dad,

It sure is good to get mail again. I got your letter dated March 17th today. It was the first letter I got. I also got one from Sylvia today.

I am glad you were able to get another car. It sure does sound nice. What happened to the Valiant?

I am now Chief Clerk in the Letters of Sympathy Division. The guy that was breaking me in went on R&R today. He will be going home for good in about 27 days, so I will have to learn my job well. I sure do have a lot of work, so time goes fast when I am at work. I hope we get another clerk to help me out. We all have to work Sundays and Saturdays. We do not get a day off. We do get one and a half hours off for lunch. We usually lay in the sun and write letters. We also have four guys that work all night. There are at least two guys in the office 24 hours of the day. I have been playing softball and volleyball to keep in shape. It is good to run around after working in the office all day. I like this job, for it gives me good experience. I have never had an office job and it helps me since I am in Business. My typing speed is getting better and better.

Last night, we had a live band and they were pretty good. Today is Sunday and we don't have to be to work until 8:30 AM. I didn't get a chance to go to church today because I had CQ. I had to answer the phones in the office while the rest of the guys went to eat. I sure do have a lot to thank God for and even though I didn't get to go to church today, he was in my thoughts.

Mother, could you please send me a pair of shoes or loafers, one pair of socks (any color) and one short sleeve shirt. I want to get these things ready for R&R when I come home.

Take care and God be with you.

Love,
Tom

March 22, 1971

Dear Mother and Dad,

We get mail twice a day here, and your letters got mixed up.

I found out about your accident mother and what happened to the Valiant in the night mail. I am glad you were not hurt too badly. It must've been an experience. I am pleased to hear that you got a VW, because they are pretty dependable. God sure has been good to us and we have a lot to thank him for.

I am really busy now, but I enjoy the work. Time goes by so fast when you are busy.

I have bunker guard tonight with three of my new friends: Steve Pittman from Arkansas. He is a slow talker. Jacobs from California, he raises pigeons. And Henry Brasseux from Louisiana. Henry lives just outside of Fort Polk. The mosquitoes are bad here, so could you send me some insect repellent.

They are worse up here than down south. It is getting warmer here and the hot summer will be here soon.

I still have some film snapshots in my camera, but I will take some slides pretty soon.

My days are the same, so little to talk about.

Take care!

Love,
Tom

On a trail in the Central Highlands, February 1971

March 25, 1971

Dear Mother and Dad,

I got letter #27 today. It sure is good to hear what is happening at home and get some mail again.

I do not know what Pat's full address is, could you send it to me? I wrote once to him, but I had to send it to the company's headquarters. I hope he makes out okay. It sounds as though the fighting is picking up down there. Pat and I are good friends, and I think about him a lot.

Yes, I have seen some fighting, but I did not want to write about it because I did not want you to worry. I was never in a big fire fight, and none of my friends were killed. I had a lot of sleepless nights, and many times I prayed myself to sleep. It is always the unexpected event that makes you worry. Sometimes I wonder what I would have done if I had not known Christ.

My work is keeping me busy. The Laos operation does not help matters any.

Could you send me one pair of my khakis. I cannot remember if I have my PFC stripe sewn on one pair or not. Also send my green dress cap. I may need them to get home for my 14-day leave. Not much more to report.

Take care!

Love,
Tom

P.S. Jim McBride is getting married to Sandy this month!!

P.S.S. Also, send me one white T-shirt.

March 30, 1971

Dear Mother and Dad,

Sorry I haven't written sooner, but I have been very busy. I have to work nights sometimes to catch up with my work. We are really busy. The weather is getting nice, and this is when the troops and supplies are on the move. The extra details we have to do also keep us jumping. Last night I had all night guard on bunker line, and today I have day guard until 5:00 PM. At least this way I can get some letters written and work on getting a tan. However, I lose out on a lot of sleep.

I got a nice letter from Mr. Ross, the economics teacher at Bloomsburg. He helped me with the swimming team last year. The swimming team had a 10 to 5 record and placed third in states. Dave Gibas won the 50 and 100 freestyle for the second time. He also placed 1st in the 50 and third in the 100 in the NAIA championships. I imagine Coach McLaughlin is very happy. This qualifies Dave as an all American college swimmer. Quite an honor for Bloomsburg.

We had quite a few fellows leaving for the states in my office. Four of them will finish their tour in about 20 days. This includes our captain. We had a little party for them a few days ago. Oh well, this way I can make new friends.

I plan on coming home in the middle of June for my 14-day leave. There is a little red tape involved, so I can't give exact dates. I will have to buy my ticket at least one month in advance, so that I am assured a seat.

Sylvia mentioned it was good to see you and talk. I was glad to hear that Mary Beck made the trip with her. [Alice Beck was in his class in Waverly. Her sister, Mary Beck, worked in the same school as Sylvia outside of Shamokin; she went to Bloomsburg for teaching, too. She probably went along to see her folks in Waverly.] They get along pretty well.

I have the roll of slides in my camera now, so if you want to send me another roll, I will keep taking pictures.

Phu Bai means the land of the dead, and there are graves all over this area. They are beautifully decorated with pillars, cement walls and figures of dragons surrounding the grave. I will take some pictures of them and the mountains. Other than this, I will not have much to take pictures of. I do not get around as much as I did when I was a "grunt," but that all goes along with the benefits of having a rear job.

Look over the PX catalog, and when I get home from my leave, I will take orders. I was thinking about taking a course in computers by mail while I am over here, but my time is limited, so I decided not to. I will try to read more, but will have to discipline myself. Not much more to write. I am fine and in good health.

Take care!

Love,
Tom

April 2, 1971

Dear Mother and Dad,

I got your care package today and thank you very much. I sure can use it.

My work is coming along okay. Cooper, the other clerk, came back from Hawaii yesterday. I just about have my work caught up, thanks to him. He will be getting out of the Army in 20 days. Then I will be by myself again. I hope the fighting stops, then my job wouldn't be so hard.

Do you plan on enlarging your garden this year? It sounds as though you plan on having a larger garden this year. I know Dad likes to work in the garden and it is good for both of you to get some sunshine after being inside all day. I lay in the sun for about 45 minutes every day.

Thanks for helping Sylvia out with our business papers and the car situation. I know she appreciated it. There really isn't much to write about, for I do the same thing every day.

We have already had our rainy season up here. It starts around October and goes to February. The rainy season should be starting where I was when I was with the Cav. Maybe I will miss the rainy season all together.

Take care

Love,
Tom

P.S. By the way, our chaplain is real good. He jokes and gives a good sermon instead of reading it.

April 7, 1971

Dear Mother and Dad,

Today was letter day. I got a surprise letter from my buddy that came over here with me, but then I didn't hear where he went. Pete gave him my address. I think I mentioned him Bob Stevens from Vermont. Pat and I, Pete, Ed and Bob went to Baton Rouge together while at Polk. Well anyway, he is an educational NCO at an MP stockade in Danang. Danang is only 50 miles from me. He likes his job and is thankful that he is not in the bush. He is getting married when he goes home in July for his 14-day leave. It sure was good to hear from him.

I will look into the allotment that I made out to you. With all this moving around that I have been doing, they probably got it mixed up.

I plan on coming home June 14th, give or take a few days. I hope I will be able to get a plane ticket. The wallpaper is pretty, and I imagine Dad is doing a fine job with it.

I also got a letter from Mr. and Mrs. Sharpless, the people we rented our apartment from in Bloomsburg. They are fine and plan a trip to Georgia this summer.

I am fine my work is going slow now, which is very good.

I did write to Gladys and told her about my new job, but I haven't heard from her. [Gladys is his grandfather's wife.] I got a letter from the Hoers [Cheryl and George], Smiths [Aunt Joe and Uncle

Norm], and Longs. I guess I do pretty good on mail call.

That's it for now. Take care and God be with you.

Love,
Tom

April 9, 1971

Dear Mother and Dad,

Not much new here! My work is slow right now, but that is good.

We played touch football tonight. It is good to run around and get some sun after a hard day in the office.

Steve Pittman from Arkansas and I are pretty good friends. He is a Baptist, and we have been going to church together. Thursday night, we went to communion, and they had a little service. Today was good Friday, and we had a guest speaker. He was okay, but not much zip in the sermon. Our regular chaplain sure can give a good sermon. Easter Sunday we should have a good service.

Today it was a little cooler and made work a little more enjoyable. I think it is going to be a real hot summer.

The food in the mess hall is really bad. I was sick in the beginning of the week, but I feel fine now. Your care packages and Sylvia's keep me filled up at night.

No, we did not get a chance to see any of the village near us. We do not get much time off and a lot of the villages are off-limits to military personnel. They still have VC in the villages up here.

That's about it for another day.

The other guy from California that raises pigeons is Dan Jacobs. He is going home for his 14-day leave on April 25.

Take care!

Love,
Tom

April 10, 1971

Dear Mother and Dad,

This is just a quick note! I got your package today and thanks a lot the clothes are perfect. Do not send my khakis. One of my buddies gave me a set. I hope this letter gets to you in time before you send them.

I checked on your allotment, and they did not have one for you so I made out another one. This time I made it out for $50. It will start in May.

I got half a day off today. We will be getting one every week now. It is good just to relax.

I got a nice card from Uncle Dick and Aunt Polly. They are well. I hope Henry is feeling better. He must have gone through a lot.

Take care, I will write more later.

Love,
Tom

P.S. Thanks for the package and Easter goodies.

With dog in front of reinforced shelter, May 1971

April 13, 1971

Dear Mother and Dad,

I thought maybe my letter would not get to you in time, but that is okay. I will be wearing my khakis home when I come home for my 14-day leave. I will be looking for them in the mail and thanks for sending them. I can get a green hat, shoes, and belt over here. I forgot I wore my green cap to Fort Lewis. No, I did not have another one.

I hope to get my ticket pretty soon. Then I can tell you all when I will be coming home. Maybe I will find out something tomorrow.

I got a care package from Sylvia today. She made whoopie pies, and they are very good.

I am on office CQ tonight with Steve Pittman. We will get off at 7:30 PM. Then we will probably watch a movie. Steve is a good kid (23 years old), and I enjoy working with him.

Sylvia sent me a picture of Jim's wedding, and it looked very nice. I am happy for the both of them. They are good people, and Sylvia and I get along real well with them. She also sent me a picture of her dog, Taffy. Taffy looks like a cute dog, but the color in the picture was a little off.

I got a nice letter from Pat today. I wrote him earlier in the week and told him what I was up to. He sounds in good spirits and thinks his unit will be getting out of the bush soon. Pat and I are good friends, and I like to keep in touch with him. His parents are also nice people.

I really do not know what I want to do on my 14-day leave. Sylvia and I will have to discuss it. There is so much I want to do and so little time to do it.

My work is okay, but I am busy most of the time, which makes the days go by fast. It doesn't seem possible that I have been over here 4 months now. It is going to be good to get back to the states.

Not much more to report. I am fine and in good health.

Take care!

Love,
Tom

April 16, 1971

Dear Mother and Dad,

I got a nice care package from Uncle Norm and Aunt Joe. I really appreciated it.

Captain Peterson, my boss, and I went to Camp Eagle today. I picked up my plane ticket for my 14-day leave. It is all set now, if the Army doesn't do anything foolish. I will leave Saigon at 11:00 AM 14th of June and fly into Chicago at 6:30 PM the 14th June. Where I will be flying into to complete the rest of my trip is undecided as of yet. Sylvia and I will have to talk it over and see what would be the best for her. I am excited about going home and being back in the world for about 11 or 12 days.

My work is slow right now, but that is good. I am fine and the weather has been just beautiful the past few days.

The other day I went before the Sergeant Major. I am up for promotion to spec-4. I will probably get it in about two or three months, I hope. It means an additional $50 per month.

I hope everyone is fine at home.

Sylvia said that she and her mother were going to see you on Easter. I hope it was a nice day. She also said you were going to Trevorton on May 1st.

I really think about you and Dad a lot. I wish I could help Dad paint the house and work in the garden.

I hope the four of us can do some traveling together when I get out of the Army, for Sylvia and I enjoy your company. I know how much you and Dad enjoy traveling, and we could be some good company for you.

Take care and God bless you.

Love,
Tom

P.S. Dad, what do you think about the SRT101 Minolta camera? They have it in the PX catalog. You can change lenses on it. A friend of mine has one over here and he likes it. Maybe you would like one?

I will be taking orders in the PX catalog when I get home.

April 25, 1971

Dear Mother and Dad,

I got your letter yesterday and also the enclosed alumni news. I appreciate it very much.

It looks as though Bloomsburg [University] is really progressing, with new buildings and new courses of study. Like anything, it has to keep up with progress.

I am sorry my remaining pictures from that roll did not turn out so well. I hope it is nothing wrong with the camera. I will be sending another roll of film shortly. Could you please send me some film, for I will be taking a lot of pictures when I come home in June, and I can't get the film over here.

I got my military clothes that you sent me. I will have to press them off for the trip home.

The weather here is extremely hot. You can't do much, for you sweat too much.

I had perimeter guard last night, so I had the morning off. It is impossible to sleep in the daytime.

Friday we had a nice party for two of the guys in the office and the captain who are going home. We started the night off with a softball game, which we won. Then we ate hamburgers and drank soda and listened to music. After that, we went to the officer's basketball court and played basketball. It was an enjoyable time.

My work is increasing now, for we are going into the Ashau Valley. I hope we get out of there soon.

Take care!

Love,
Tom

April 27, 1971

Dear Mother and Dad,

I got two letters from you and three from Sylvia. The mail sure does get messed up every once in a while. I also got a letter from Mr. Berry and Mrs. Allmand. Everyone seems to be fine down that way.

I am glad you sent me the clippings of Waverly's track team. It is good to hear how they are doing. They seem to have a pretty good team this year. Bloomsburg's track team is doing real well this year, and they should place high in the state championships.

My work has picked up, and once again I am behind. I wish they would get a few more people in the office. It is hard to do a good job if you are pushed to get it done. At least I am busy and the time goes fast.

Could you please send me the Old Testament part of the Bible? I tried to read my little New Testament every night, and I am almost finished with it. I promised myself that I would read the Bible while I was over here.

I haven't been doing much exercise lately, because it has been too hot. They shut the water off, so it is hard to shower after you are all sweated up. I guess water is a problem over here during the hot season.

One of my friends over here has gone home for his 14-day leave. He lives in California, so it is easier for him to get home. I imagine he is living it up right now.

My captain has left and so did my other buddy, Dan. Dan is out of the Army now. He extended his time over here, so he could get out of the Army five months earlier. I think he extended 60 days.

Well, not much more to report.

Take care!

Love,
Tom

April 28, 1971

Dear Mother,

I want to wish you a happy Mother's Day. I am sorry I was unable to send you a card, for they are hard to locate over here.

I got a letter from you and Sylvia today. It is good to hear from home and also hear how Waverly's track team is doing.

I do not know if I told you, but Sylvia and I decided to pay off our car. It should be all ours now. We had a little extra money saved up, so we decided to pay it off now and not pay for all that interest.

I am glad Dad is thinking about getting another camera. He takes such good pictures, and I think he would like a better camera. My catalog has some pretty good ones at a good price. I have been thinking about buying Dad's camera that he has now, if he decides to get another one. We will talk about it when I get home.

It looks like I will try to fly into Syracuse or Elmira when I come home for my leave. You are invited to come with Sylvia and pick me up. I know it depends on when it is and at what time. Sylvia will be on vacation, but you and Dad will have to work. I will probably call from Chicago and let you know where I can get a flight to and what will happen.

My work is keeping me busy, but time goes fast.

I will send my film to be processed tomorrow.

Have a nice Mother's Day, for you are in my thoughts and prayers.

Love,
Tom

May 3, 1971

Dear Mother and Dad,

I am on office CQ tonight, so while I have the time I decided to write you.

I was a little disturbed last night when I found out that someone stole my camera. I guess you just can't trust anyone. I am glad I didn't buy an expensive camera for then I would be out a lot of money. My Instamatic did take good pictures, and I would like it back, but there are no clues. They also got my blanket and poncho that I use while on guard. Oh well, there wasn't much to take pictures of here anyway. No need in you sending me any film.

I have my work caught up to date, and it looks like maybe our work will slow down. I hope so.

We had typhoon warnings last night, so they canceled movie. It has been raining for two days off and on. It didn't do much last night, and I did get a good night's sleep.

I do not know if I mentioned it, but Pete Abatielle got a rear job at Bien Hoa. He is working in the personnel management section. I hope he can help Pat get out of the bush.

We have a new chaplain, and he is a Southern Baptist. He is pretty good and puts feeling into his sermons. I also had a chance to receive communion last Sunday, which I did. They have communion once every first of the month like we do.

I don't think I mentioned it before, but quite a few of the guys that are getting killed over here are Baptists. I don't know what that means, maybe there are more Baptists than one thinks. It was just something I noticed.

Not much more to report. I am fine and feel good.

Take care!

Love,
Tom

Dear Mother and Dad,

I found my camera! When I came off guard, I was pretty sleepy and I must have put my camera in the bottom of my locker. So anyway, I have a camera, and I feel much better. I could use some more film for when I come home next month.

I went to the movies after office CQ last night and saw a comedy by Don Knox. They mixed up the reels and showed the second real first and the first real second. What a way to watch a movie.

I just wanted to let you know that I found my camera!

Love,
Tom

May 5, 1971

Hi Mother and Dad,

The mail sure is a problem lately. For two days I don't get any mail, and then the third day I get about six letters. I got two letters from you yesterday. It is always good to get mail from home.

There is no fear about me losing my job when I come home for my 14-day leave. It will still be here waiting for me when I get back. I am in the process of showing John Dyer how to do my job. He came in from the field, and now he has a rear job. He is from Tennessee and a very nice guy.

I am writing this letter at the office, so you can see my work is tapering off. I have everything under control.

Our new commanding officer is really nice. He is from New York City, what a small world.

Yes, I think Steve Pittman is in the picture that I took while sunbathing on top of the bunker. I will have to tell you about the pictures when I come home.

I still haven't decided where I will fly into. I hope I can get to our Elmira or Syracuse. I will have to wait until I get to Chicago to find out. [He was flying standby for domestic flights.]

Not much more happening.

Take care.

Love,
Tom

May 10, 1971

Dear Dad,

I was talking with Jacob, one of my buddies, who came back from a 14-day leave from the states. He said his plane made a stop in Japan to refuel the plane. While he was there, he bought a zoom lens for his camera. I was thinking if you are considering buying another camera, I could get it for you on my way home. The prices are the same, if not less, than the ones in the PX catalog. You could also save money on postage and insurance. It is worth a try.

Let me know what camera you would like. Also, send me a $100 postal money order, and I will make up the difference. If you want any lenses, just let me know.

Let me know what you think.

Love,
Tom

May 10, 1971

Dear Mother and Dad,

Today is Monday, and I have the whole day off. We will get about two days off every month now, since our work is tapering off. I didn't intend to spend my day off in bed, so I got up and did my wash and am cleaning up my AO (Area of operation). It is only 8:30 AM, and the sun is shining brightly. It is going to be a hot day again. I intend to write letters and catch up on some of my reading.

I have decided to take an Army correspondence course. I am taking the course to change my MOS. I still have an infantry MOS, but when I make spec-4, they will promote me to clerk typist. If I go back to the states with a clerk typist MOS, they will not give me a good job while I finish up my time in the Army. So I decided to take this course to change my MOS to personnel specialist. This correspondence course will also give me points for the spec-5 board.

I have been thinking about the idea of going into administration work either at the high school or college level. A lot of it is going to depend on what type of jobs are available when I am looking for a good job. After I get my Masters, maybe I will have a little more bargaining power. With unemployment so high, maybe I could just settle for a teaching job. We will see.

Sometimes I wish I were back out in the bush. Physically, it was much better for me, instead of sitting behind a desk. But, I am thankful for this job and at least I had the opportunity to see what the bush was like.

The other day we had a guy commit suicide. The sympathy letters are going to be hard to write. It is a shame that a person gets so disturbed and takes his own life.

Two important dates are coming up for me, which we will have to celebrate when I come home for leave. Take care!

Love,
Tom

May 11, 1971

Dear Mother and Dad,

I don't remember if I thanked you for the care package or not. I did receive it and thanks a lot. It had film in it also.

I called the R&R Center to find out about the flights out of Chicago. I was told that it would be difficult to fly to Syracuse on the 14th of June. The next best thing looks like Philadelphia. I am supposed to fly into Chicago at 6:30 PM. There is a flight leaving Chicago at 8:30 PM and gets into Philadelphia at 10:40 PM. This looks the best to me. I hope Sylvia doesn't mind making the trip to Philadelphia. The airport should be easy to find. It doesn't look like you will be able to meet me at the airport. We will drive up home the next day. I guess one more day doesn't make that much difference. Sylvia and I will stay at a motel that night and then drive to Waverly the next day. I just wrote Sylvia today about these arrangements. This is our tentative schedule.

I asked Sylvia to get my driver's license from home, so I can help her drive. My license is in my wallet in John's room in the top drawer of the dresser. Could you see that Sylvia gets my license? I hate to drive without it.

It was really hot today. Just sitting at my desk the sweat poured out of me. I didn't sleep well last night either, because of the heat.

Do we still have that little fan, Frosty? I would like to bring it back with me. Most of the other guys have a fan and it sure does help. I need something.

Is there anything you would like me to bring home to you for a souvenir?

I am really moving up in the world. I got my own phone yesterday.

I hope my arrangements work out OK. I will just have to wait and see.

Take care!

Love,
Tom

May 16, 1971

Dear Mother and Dad,

I got your letter yesterday, and I sure enjoyed it. It sounded as though you enjoyed your Mother's Day.

Not much is new with me. My work has slowed down, so I can relax a little more.

Yesterday I had R&U, repair and upkeep. We dug a few ditches and laid out some pipes. I got some sun on my arms, and it felt good to work in the sun.

I guess I will have to be careful with the pictures I take from now on. My picture taking is slowing down, for there is little to take pictures of. I will still take pictures of my friends, but not much else.

I have been thinking about my arrangements when I fly into Philadelphia. I told Sylvia that I would like to have her drive down during the day, so she could be down there and take her time. I thought I would call you from the Chicago airport and let you know what flight and what time I will arrive in Philadelphia. I thought Sylvia could call you from Philadelphia and get the information so she could meet me at the airport. There are two flights out of Chicago: one at 7:15 PM and the other at 8:30 PM. I should be on one of them. I hope everything works out.

That's about it.

Take care.

Love,
Tom

May 17, 1971

Dear Mother and Dad,

Thank you for your care package. I received it yesterday, and I also got your letter today number 48. I have started reading Pleasant Valley. It looks good so far. I have also started reading the Old Testament in my Bible, for I finished the New Testament. Thank you also for my flashlight. I sure can use it. We will celebrate my birthday and anniversary when I come home.

I got a letter from Sylvia today and I think we have come to a conclusion on my flight home. I will get a flight out of Chicago at 7:40 PM and fly to Allentown where Sylvia will meet me at 10:23 PM. I told Sylvia to call United Airway and reserve me a seat on that flight, then I will buy my ticket when I get to Chicago at 6:30 PM. This way she will know for sure that I am on the flight. I think Allentown is the best bet. Ron Bishop's parents live in Allentown and his parents can give Sylvia directions to get to the airport. Ron was the one that married Sue [Sylvia's sister] and lives in Wisconsin. We will drive up home the following day.

I had a pleasant surprise today. Bob Stevens called me from Denang. He is excited about going home in July to get married. He wanted to know if I have heard from Pat, Pete, or Ed. Since I haven't, I didn't have much to tell him. It sure was good to talk with him.

Yes, Sylvia knows of my renewed interest in the Bible. This situation has also given her a renewed interest in God. She asked me if we could have family Bible reading in our home. God has brought us together from two different backgrounds, yet we have a lot in common. We will work it out together with God's help.

That sure is a nice write-up John got in the paper. He must be extremely busy.

I also got a birthday card and letter from Uncle Herb and Louise. It was good to hear from them.

Waverly had a pretty good track team this year. Their times were better than when I was in high school.

I do not understand why New York State has so many problems with money. I think this cut back in the number of teachers at Waverly will hurt their educational program. Money sure is getting tight.

I hope Dad will let me know if he wants another camera. I am sure I can pick him up a good camera at a good price in Japan.

That's it for another day.

Take care!

Love,
Tom

May 21, 1971

Dear Mother and Dad,

Today was a pretty good day. My work was slow, so I did some reading. I finished the book you sent me, Pleasant Valley. It was different from most books I have read, yet quite good. He sure had a fulfilling life and such a beautiful place. The animals he talked about were very interesting and most had human characteristics. I enjoyed the book. Thank you.

Besides your letter today, I also got two from Sylvia and one from Donna and Charles. It would be nice if Charles could get that teaching job in Ohio.

The other day I got a card and letter from Uncle Herb and Louise. It was good to hear from them.

John's promotion sounds pretty good, and I imagine he is happy about it.

I hope you enjoyed your visit to Trevorton and the weather cooperated with you.

I think I told you that I am flying into Allentown. I am still going to try to call you from Chicago and let you know how I am making out. Then Sylvia can call you to make sure I made the flight to Allentown. I think everything will work out.

If Dad is thinking about getting another camera, I think I will buy Dad's old one. I really do not have the camera fever yet, but I do plan to travel and maybe I should have a good camera. We will see.

Not much to take pictures of where I am, but maybe I can get some on my way home.

Take care.

Love,
Tom

May 22, 1971

Dear Mother and Dad,

I got your letter that dad wrote on the 16th and also the money orders. I also got a letter from Sylvia today. It sounds as though you had a nice time and enjoyed your visit. I hope I get a chance to see the mines in Trevorton. I was glad the weather cooperated with you for a while. I ran again today during my lunch hour, and although my legs are sore the experience is good for me.

Tomorrow is my wedding anniversary and also my day off. I plan on getting my shots tomorrow so I will be all set to come home.

I am pleased that you decided to buy the Minolta SR-T101, Dad. Two of my friends have the same camera and they really enjoy it. The telephoto and wide angle lens will also add a new phase in your picture taking. I have enough money to get everything you want, and I am sure we can save money this way.

My work is picking up, but with two of us working on it, it is not too bad. I sure wish this war would end, so I wouldn't have to send any more sympathy letters.

Rick Bastwich from Waverly must be a good sprinter. And 9.8 for high school is outstanding. Do you think he would like to go to Bloomsburg?

Take care —23 days and counting.

Love,
Tom

May 23, 1971

Dear Mother and Dad,

Today is my wedding anniversary, but it seems like any other day. It is another hot one and it was unbearable in church today.

Rick Rathbone and I went to second brigade to get my shots and he had to get some clothes altered. The shots didn't hurt, but my arms are a little sore right now. I tried to sleep a little this afternoon, but it was too hot.

There isn't any place over here where I can buy a fan. I intend to get one while I am home and carry it back. They are not that heavy.

I sent three pairs of sandals home today. I hope they get there. One is for me, one for Sylvia and one for John. They are called Ho Chi Min sandals.

I also got my care package that you sent and also the film inside. I haven't been taking many pictures, but I hope to on my way home.

Well I guess that is about all.

Take care!

Love,
Tom

May 31, 1971

Dear Mother and Dad,

I did real well at mail call today: two letters from you and two from Sylvia. I guess that explains why I haven't gotten much mail in the past few days.

My orders for SP4 came today, so I will be getting an additional $50 a month. You should have or should be getting your $50 from the Army for this month.

I guess my plans are all set to go home. Sylvia made a reservation for me to fly from Chicago to Allentown. I will still call you when I get to Chicago to let you know how I am making out.

Sylvia sent me the clipping from her paper about the coaches, but yours had more in it about the situation. I think I know what the problems are, since I worked around them while I was coaching last year. I really didn't believe that the coaches would go so far as to resign, but they did.

The weather is still hot and I guess it is going to be hotter.

I guess Sylvia told you that she bought a bedroom suite. We wanted one and she got a good deal on it, so I am glad she bought it, instead of waiting for me to come home. Her mother is also buying furniture for her living room.

My work is slowing down, so I have a little more time to read.

We had a rocket attack last night and since I was on the reaction force I had to get up, get dressed, and fall out in my gear. Fortunately, we did not have a ground attack, so I got to go back to bed at 2 AM. Such is the life of a soldier at Phu Bai.

Take care!

Love,
Tom

June 2, 1971

Dear Mother and Dad,

I got a nice letter from Pat today. He is still in the bush, but goes back to fire base [*illegible*] every 15 days. I think he is sick of the weather and I know I would get tired of being wet all the time. He said his grandfather died and he had to go to Bien Hoa and call home. While he was there, he got a chance to see Pete. He also mentioned that his mother and uncle have three horses now.

The weekend of the 18th and 19th of June, Sylvia and I will be home. It doesn't matter what we do. I would like to see some of the relatives, but I do not care if they want to come to our house or we go visit them. Whatever you decide is fine with us, or whatever would be the best for them. Sylvia and I do not have definite plans for my leave, we will just do things as they come along. I would like to see you Tuesday night the 15th, but it depends on the time. We will call and let you know what we decide.

My work is picking up, but it comes in spurts. Last month was the slowest month in a long time. I keep a lot of the statistics and they were low for the month of May.

I will be home soon!

Take care.

Love,
Tom

June 8, 1971

Dear Mother and Dad,

I received two letters from you yesterday, it is always good to hear from you.

I am glad you received the $50 and sent rotary the money. I got the letters on the same day, so I didn't have to go to finance and check on the allotment. I will try to talk with Dr. Schaeffer when I get home.

Most likely I will beat this letter home, but I felt like writing.

I have been working on my correspondence course from the Army and it gives me a better idea just how the Army is run. It is mostly cut and dried, but I am learning.

Glad to hear that Aunt Polly and Uncle Dick were able to come down for Memorial weekend. It is good to get out and do different things.

We had a softball game tonight, which we lost. I enjoyed playing, even if we don't win.

See you soon! If not already!

Love,
Tom

June 8, 1971

Dear Mother and Dad,

I sure enjoyed my 14-day leave and it was good to be home for a while. Thank you for making my stay enjoyable. Sylvia and I really enjoyed the parade and it was nice that we could all be together.

My flight back was OK, but long. I mostly slept and read my Reader's Digest.

I called Sue and Ron while I was in Chicago. It was good to talk with them.

I spent the night of the 27th at Camp Alpha. I was extremely tired, so I went to bed early. I got a flight out of Saigon at noon and got to Bien Hoa by two. Captain Miller gave me the afternoon off, so I unpacked and straightened up my AO.

Tonight we got a set of horseshoes, so we played until it got dark.

Casualties are slow right now, that is good.

Not much else to report. I will write later.

Love,
Tom

Tour of the city of Hue, June 1971

July 3, 1971

Dear Mother and Dad,

Sorry I haven't written sooner, but I have been busy. I have received two letters from you since I returned.

I was sorry to hear about Pat. It must be rough fighting where he is now. I guess the Lord was looking after him when he got malaria and was not with his unit when they got hit. I hope they do not send him back to the field. Most of the malaria cases over here are not too serious. They have the medicine and facilities over here to treat it, so it should not be too bad.

Today 15 other fellows and myself went on a tour of the city of Hue. We went in the back of a truck and I was elected to carry a rifle and ammo, which we did not need. We visited many temples and Buddhist structures and traveled down the streets of Hue. I took about two rolls of film and I hope they came out. It was hot and I think I got more sun than I should have.

We also had a party tonight for one of the fellows in the office that finished his tour and is out of the Army. He extended to get out of the army earlier. He left yesterday, so he missed his own party, but I don't think he minds too much.

We have a set of horseshoes and we play every afternoon. I am getting pretty good, but I do have some competition.

Captain Miller is leaving in about 20 days, so we now have a new officer in charge of the office. He seems like a nice guy and he is from North Carolina.

Sylvia and I have been discussing about me extending over here, so I can get out of the Army earlier. This way I may be able to go to school starting in February. What are your ideas on the subject? I hate to miss another Christmas, but I would be out of the Army in February. If [*illegible*] go up, I do not think I will extend and they look like they may. It all depends on a lot of things, so I think I will wait and see how things work out. We were just discussing it.

I got my fan that I sent the other day. I sure use it a lot, for it is quite hot.

I seem to have a rash between my legs. I did not have it while I was home. Could you send some medicine or something for it? It sure itches.

Work in the office is pretty slow now, but some days are worse than others. Today we had four men missing. Deaths seem to be going down.

I ordered your wide angle lens, Dad and I had to send it air mail registered. Since I also had to send Sylvia's ring air mail, I had it sent home with your lens. Let Sylvia know when it arrives. The salad bowl and binoculars I had sent to her by parcel post.

I guess that is all.

Take care!

Love,
Tom

July 8, 1971

Dear Mother and Dad,

Well, I made it through my first typhoon. It hit us on the 6th of July. We went to work, but the lights went out at 8:45 AM, so we all went back to our barracks and put some sandbags on our roof. Sections of the roof flew off the building where I work, but not around the casualty branch. At around 4:00 PM the storm was really bad, high winds and rain. It settled down and we all got a good night's sleep. There was not too much damage and no reports from anyone getting hurt. My barracks held up pretty good and no damage, just a little wet.

It is 7:15 AM and I just got back from guard duty. Yesterday turned out to be a nice day, so the typhoon is just something of the past.

Yesterday we had an important volleyball game which we lost. It was a good game and we all enjoyed playing; win or lose.

Thank you for my care package. I received it yesterday. I hope to get my film sent out soon from the tour of Hue.

Work is okay. Right now it is slow, but that is good.

Not much more to report.

Take care!

Love,
Tom

July 11, 1971

Dear Mother and Dad,

Sorry I haven't written sooner, but the time seems to go by pretty fast. I have received your letters and appreciate you keeping me informed about Pat. I am glad Pat is out of the hospital and that he is going to sniper school.

Dad is probably painting the house by now. I hope you are enjoying your vacation.

I have some good news. On 18 July, I am going before the E5 board. It is a special board, because Captain Miller feels I am doing a good job and should be SP5, even though I do not have the necessary months in the Army or time in grade SP4. There are five men on the board and they ask you a variety of questions. You are scored on a point basis and the ones with the most points will get E5. There are not many allocations for each five, so it is going to be real hard to make it. I really appreciate the opportunity to go before the board. I would like to make it, but I know it will be hard. I can try again if I don't make it this time. I already have a few points going into the board, with my civilian schooling, combat infantry badge that I earned with the 1st Cav division, and maybe a few points for my correspondence course I am taking. It will be interesting how it will work out.

I was on repair and upkeep today. All I did was guard the Vietnamese women while they worked.

My work is fine; slow, but that is good.

Everything is fine with me.

Take care!

Love,
Tom

July 13, 1971

Dear Mother and Dad,

It has been raining steady for the past two days. I have never seen anything like it before.

I had interior guard last night and I am glad we have wetsuits (pants and jacket). The only thing that got wet were my feet. I finally got off guard at 2 AM and it didn't take me long to fall asleep. George McClead, a buddy of mine, he works in the same branch, he was on guard, too, so we talked and it made the time go faster. He is from Paducah, Kentucky and has a 19-month old baby boy. We have a lot in common and he is a good guy.

Tonight I have bunker guard with Jacobs, my friend from California. We got the same bunker, so we will probably talk a lot tonight.

I have decided not to extend over here. It is just not worth it. It would be nice to get out of the Army early, but there are too many details over here and I just plain don't like it here, especially for two more months if I extend.

I am glad to hear you got to go on a little vacation. Yes, I do remember when we went to [*illegible*] Dam. It was a while ago. I imagine it has changed a lot since we were there.

Everything is fine with me. No complaints. Work is slow, but of course that is good.

I hope Mr. Ransom is feeling better. He sure is having a lot of problems.

Take care!

Love,
Tom

P.S. We were supposed to have another typhoon, but it only rained.

July 19, 1971

Dear Mother and Dad,

I got two letters from you today. After not getting any mail for three days it was good to get some letters today.

It sounds as though you enjoyed your trip to Cooperstown. I am glad you were able to go.

Not much is happening here. I guess I will not be going before the E5 board this month, because they did not get our names in soon enough. There is always next month. I am not sure if I will go up next month or not. I may have to wait until September.

I still have my rash. Some days are better than others. I think I will go to the aid station tomorrow and see what they say.

I have bunker guard tonight. It will not be too bad, for Jacobs and I have the same bunker. Jacobs is starting to work nights, so I will not see as much of him.

Today is Tuesday, 20 July 1971. I was not able to finish this letter last night.

I went to the aid station today after guard and I have a bad rash. They gave me surgical soap and some ointment. I have to go back in seven days.

Not much really happening here. The casualties are down, so that is good. We have been playing horseshoes after work. It is fun and everyone enjoys it.

Captain Miller finished his time over here and he left last week. We now have a second lieutenant in charge of the office. He seems to be a nice guy and we all get along good with him.

Take care!

Love,
Tom

July 22, 1971

Dear Mother and Dad,

Thank you for the care package. We ate some of the peanut butter last night for a snack and I made some Start last night. I drink about a quart of juice each day. Now that I have a container for it, it is easier to mix.

The movie last night was a rerun, so we just sat around and talked. I was thinking about buying a stereo set up and we were talking about them and comparing prices.

Jacobs and I had KP yesterday and we were dining room orderlies. It didn't hurt me any, but I don't care for KP.

I had a letter from Pat a few days ago. He seemed to be in good spirits. He doesn't stay in the bush as long as he did before and they have a base camp that they use as headquarters. Concerning the python he has as a pet, we have everything over here. Besides snakes, we have tigers, bears, and elephants. You name it and I think we have it.

My rash is getting a little better, but no big signs of improvement. The itching has stopped. I am still putting on the medicine that the doctor gave me. It sure is uncomfortable.

Glad to hear that you got a postcard from Ralph Taylor and his wife. I would also like to visit that section of the country. Maybe when we visit Sue and Ron. Casualties our way down and I hope they stay that way. I have been reading on my slow days, so I am not just wasting my time.

Pete Abatielle and Bob Stevens are home on leave right now. Bob is getting married while he is home.

I guess that is all for now.

Take care and God be with you.

Love,
Tom

July 27, 1971

Dear Mother and Dad,

I really appreciate your letters. The mail has been getting fouled up lately. One day I get five letters and the next three days I don't get any.

Sorry to hear about Mr. Ransom. He is better off this way, but it is always sad.

Nothing has been said about me going before the E-5 board, but I might this month. It doesn't matter to me when I do, for I will get a shot at it sometime before I leave here.

Now for the good news, Sylvia might be pregnant. It is too early to tell for sure, but it is highly possible. We discussed the possibility before I got home and we decided to let it run its natural course. We were careful, but you know how things like this happen. We are both happy and I think we are ready for a family. I know it would have been better if we could have waited a few years before starting our family, but I think the Lord decided to bless us with a child now. Waiting to find out for sure is the hardest part. I wanted you to know now, for Sylvia thought it would be better if I told you and by the time we find out for sure it would have been old news. Our plans have not changed much. It depends on where I will be assigned when I get back to the states and I am sure everything will work out for the best.

The weather is still hot and I will be glad when the colder weather settles in.

I got a letter from Pat and he told me about the same thing that Mrs. M told you. [Mrs. Manderville was Pat's mom.] I worry about him sometimes, but he is rugged and can take care of himself.

Take care and God bless

Love,
Tom

P.S. In my next care package, please send me peanut butter and crackers (Jiffy and hard crackers), the soft crackers break in transporting. I am still using Start, so keep that coming.

We have had quite a few new soldiers come up here. We had about 450 process in yesterday.

Love,
Tom

July 30, 1971

Dear Dad,

I want to wish you a happy birthday. Sorry I was unable to get you a card, so a letter will have to do.

I imagine you have been pretty busy this summer, with painting the house and working in the garden. Maybe this winter you can relax.

I hope your lens comes soon, so you can use it this summer. It still might be a while, for they have a lot of orders to fill.

Not much is new with me. There is very little work to do in the office. I average about a book every two days. The 101st Division is providing support for the ARVN's and they are doing most of the fighting. Our men are on the fire bases. It will be interesting to see if the ARVN's can hold their own. Maybe we can get out of here soon. Casualties have gone way down.

The little light that you and mother sent me for my birthday is really coming In handy. They have been having trouble with the generator, so they turn the electricity off from 6:00 to 10:00 at night. The light sure helps me to get around.

Sylvia mentioned that the rust spots are starting to show through where they were patched up. Maybe when I get home in December we can go looking for cars. It seems like that is all we ever do.

I got a letter from Donna and Charles. It looks like they have a nice duplex and it is not too far from the college.

The teachers in Sylvia's district are not too happy with the administration's proposals, so they are planning not to go back to school in September. Sylvia voted for the new proposals, but 84 were against, so I guess it doesn't look good.

Take care and I hope you have a happy birthday, Dad.

Love,
Tom

August 2, 1971

Dear Mother and Dad,

It was nice to get two letters from you today. They were postmarked the 26th and 29th. No, I haven't been getting as much mail as I used to. Of course, it is mostly my fault, because I haven't written as much either.

I will not be going before the E5 board for a while. I guess they have too many E-5s now, so they do not have any openings for anyone. I would like to make it before I leave here.

Glad to see that you are still making trips on the weekends. It is always good to get out and see how things have changed.

It was nice of Ralph and Louisa to share their trip and pictures with you. They sure saw a lot of things.

We still do not know for sure if Sylvia is pregnant or not, but she has been getting sick lately and she thinks she is. She would know. It is still early for a doctor to tell.

I just got off 24-hour guard and boy am I tired. I got a day out of the office and I did get a chance to finish my book.

I got a nice letter from Sue and Ron. They had a nice vacation visiting with the relatives. They want me to order some things through the PX, which I will be glad to do.

We got another new chaplain. He has been in country about three weeks. It was good to receive communion last Sunday. The new chaplain talks to us in an informal manner and he really gets the point across.

They turned the electricity off again tonight. It is going to be warm to sleep and I wanted to get to bed early tonight. It will probably come back on again at 10 PM.

Work is about the same. I hope it stays this way.

Take care and God bless you.

Love,
Tom

August 6, 1971

Dear Mother and Dad,

How is everything with you? Things are not that good for me. Because the casualties have been on the decline and I haven't been doing much work, I am helping out in the Adjutant General's office. One of their clerks went on a 14-day leave and they needed someone, so I was drafted. I can't say that I like the work, but I guess it will not hurt me. The hours are longer and I am not familiar with the work, but I will manage. I will be working there for about two weeks. The only good point is that the office is closer to my barracks. Oh yes, and it is air-conditioned.

The weather is still hot and I will be glad for the cooler weather.

I guess they have the generator fixed, for the electricity is on now every night. The projector is broken, so the movies are over. We still play horseshoes for excitement.

My rash is all gone, but I still wash the area with surgical soap, for it cleans better.

Not much else to report.

Take care!

Love,
Tom

August 8, 1971

Dear Mother and Dad,

Happy anniversary! I have no card, but it is the thought that counts. I hope you were able to do something special on your anniversary.

I got your letter today. I was anxious to find out what you thought about Sylvia having a baby. I knew you would be happy, but news like that doesn't come every day. We are both pleased and happy and I guess all we can do now is wait.

I was glad to hear that Dad got his wide angle lens. Did Sylvia's ring come, too? Dad should just about be set in the camera line. Has John received his speakers yet?

I met Sergeant Ream yesterday. He came to the office with some Evening Times newspaper [Waverly, Sayre and Athens paper] and said I could look at them. He seems to be a nice fellow.

While I was reading the paper I noticed Lori Peterson got married. It is good to find out what happens to the kids that you graduate with.

I was pleased to hear that Carol is coming to the tournament of drums with John. Both Sylvia and I really liked Carol. She is a nice girl.

So Dad, you are at it again. If it is not painting the house, it is working on the porch. Don't work too hard! I would like to see some pictures when I get home, especially with the wide angle. Dad, keep your eyes out for some 1970 Malibu or Skylark. I am afraid we will have to get another car when I get home. I don't feel like putting too much money in on the Pontiac, since we will be getting a different car soon. The tires we need, but not a new tail pipe or muffler.

Mother, have many changes been made at work, since Mr. Ransom died? I was wondering if Bill [son of Mr. Ransom who owned the insurance company where his mom worked] will run it himself.

My new job is working out pretty good. It is still new to me, but I will manage. I had bunker guard last night with Jake and John. The time seemed to pass fast. Sylvia sent me some brownies, so we didn't go hungry.

The weather hasn't been so humid lately, so it feels a little more comfortable. Of course an air-conditioned office helps.

I have decided to buy a stereo outfit from the PX. They have some good deals and Sylvia and I like to listen to music. We really don't need it now, but we will put it to use.

Mother are you still making payments to Rotary? How much has been paid in already? Sylvia and I really appreciate the help you have given me in paying back my loan.

Take care and God bless.

Love,
Tom

August 16, 1971

Dear Mother and Dad,

Not much is happening over here. I am still working for the Adjutant General. I guess I will be here for another five or six days.

I am glad you enjoyed the tournament of drums. Maybe Sylvia and I will be able to join you next year. [This was the drum and bugle corps at Waverly High School; 6 to 8 corps got together and competed in music and marching formations— previously Tom had played drums and John had played the trumpet with the Vagabonds corps.]

I haven't heard from Pat in a while. I guess I will drop him a note. I met one of my buddies today that was with me at Fort Polk. He was in the 5th Mech [5th Mechanized Unit] and is now assigned to the 101st. It was good to talk with him.

I should be going before the E5 promotion board sometime this month. It is a hassle, but worth it if I can make it before I get out of here.

I got a nice letter from Jim McBride the other day. He will have fifteen credits toward his Doctorate when he finishes school this summer. Sandy is due in October, and Jim starts teaching in Delaware in September.

The weather has been cooler lately, which makes the days more comfortable.

I am glad you are enjoying your camera and accessories, Dad, for I got it for you to use. You can take very good pictures and I am glad I was able to get you the camera. Dad, what are the prices for an electric flash, the type you used on your first Minolta? I will have to get one to put on the camera you gave me.

Well I guess that is all I have, take care and God bless.

Love,
Tom

August 28, 1971

Dear Mother and Dad,

How is everything with you? I am still working for the Adjutant General. Tomorrow should be my last day. Major Robinson, who is my boss now, told one of the sergeants that he is pleased with my work and would like to keep me in his office. The sergeant told him that I would like to go back and work in the Casualty Branch, so I think he will let me go when the other guy gets back from his leave.

I went before the E-5 board the other day and I don't think I did very well. I am displeased with myself, for I should have done better. They asked a variety of questions, such as why did Major Lindsey change parties? What is happening in Ireland? How many keys on a typewriter? Who are the leading teams in baseball for the East and West? And a lot of Army questions pertaining to a clerk and to Army regulations. Since I am not clerk trained, I didn't know a lot of the questions. I still think I should have done better. It is the same old problem, they ask you questions that you don't know and then you get excited and when they ask you questions that you should know you get them wrong. Such is life. Maybe I will get another chance, at least I know the types of questions they ask.

George McCleade, one of my buddies, got his stateside assignment. He will be stationed someplace in the state of Washington. I think it is Vancouver, Washington. He leaves Vietnam around the middle of October. I hope I get closer to home. Most of the guys are extended to get out of the army earlier. I will take my chances back in the states.

Take care and God bless.

Love,
Tom

August 29, 1971

Dear Mother and Dad,

Not much to report, but wanted you to get a letter. Everything is fine with me. I am back working in Casualty Branch, and right now business is slow.

I got your care package the other day and really appreciated it. The peanuts were really good.

Sylvia sent me two books, which I have read: Organization of Corporations and The Generation Gap. They were really good and each had ideas that I can use in the future. I got a chance to read and really enjoy it. My Bible reading is slowing down, but I am keeping at it. Right now I am reading a book by Norman Vincent Peale; it is called the Amazing Results of Positive Thinking. Steve and George read it, and they recommended it. I have only read a few pages, and it is very interesting. I think I will benefit from reading this book. I didn't know if you had a copy of this book or not.

I had R&U detail today. It wasn't too bad.

We had a party last night for our LT. He was promoted from Second Lieutenant to First Lieutenant. We had hotdogs and hamburgers to eat and sodas to drink. The party turned out well. Jacobs and I did pretty good in horseshoes also.

Sylvia sent me a postal money order for our stereo equipment. I am glad we decided to get a set, for we will enjoy it.

I was supposed to go to Eagle Beach on Saturday, but the helicopters didn't arrive. It is a place where you can relax and go swimming. Oh well, maybe next weekend.

That's about it for now.

Take care and God bless.

Love,
Tom

August 30, 1971

Dear Mother and Dad,

I got your letter of the 23rd and 25th. I am really happy that you were able to get a trailer for your anniversary. You will enjoy it. It is good that you are camping with Ralph and Louisa, for they are good people.

You have sacrificed a lot for us, and it is now time for you to have what you want and enjoy yourself.

I also got two letters from Sylvia today. She said she really enjoyed her visit up home; Watkins Glenn, Yankee Peddler, Chemung garden food and just plain sitting on the front porch and talking. It was a welcome vacation for her. I am glad her mother was able to come up. I imagine she enjoyed Watkins Glen.

I am glad the eagle was what you wanted. Sylvia mentioned that she wanted to get one for you and she did. I'll bet it looks good over the door. The house sure is looking "early American" and I like it. With the eagle and oxen yoke out front, plus a new paint job.

I got a letter from Donna and Charles today. They sure were busy for a few days. I am glad they are settled in.

Sylvia likes her pearl ring, and I am glad I was able to get it for her. I imagine Dad really appreciated his ring that you gave him on your anniversary.

I am really glad you got your trailer, Dad. You are all set now with your trailer and camera. I really like it.

Take care,

Love,
Tom

September 3, 1971

Dear Mother and Dad,

Not much happening over here. I hope you had a good time on your camping trip over Labor Day. The radio station over here is going to play 72 hours of oldies but goodies. Steve is going to record the records so we can have some music. Other than that, not much special will happen.

I did better than I expected on the E-5 board. I am 6th on the standing list. The next time they come down with allocations for SP5, I should make it. I am really pleased.

I also got a letter of commendation from Colonel Byrne. He is the one I worked for, for two weeks. It goes in my permanent record, and I appreciated it. It stated that I did an outstanding job for him as a clerk typist. It feels good to be appreciated.

I got a letter from the Pennsylvania Senator. Sylvia wrote him to see what he could do to get me assigned close to home. He will do what he can, and we both got a nice letter.

The war casualties are on the decline, but we had four non-battle deaths this week. One guy was over here for five days and died of an overdose of heroin. The rest of the guys were shot by their own men [friendly fire]. I hope they settle down and be more careful.

I finished the book The Results of Positive Thinking. It was very good. Right now I am reading the Tough Minded Optimist Norman Peale. Could you send me the original book called the Power of Positive Thinking by Norman Peale. Steve, George, and myself are very much interested in positive thinking, and we would like to read this book.

Not much more to report!

Take care, God bless

Love,
Tom

September 8, 1971

Dear Mother and Dad,

It was so good to get two letters from you in the last two days. As usual, I got the September 4th letter before the August 31st letter. I don't know what they do with the letters. Sylvia's letters come the same way.

I was anxious to hear how your first camping trip turned out. I am glad you enjoyed yourself. I wouldn't mind trying it myself. I am really pleased that you were able to get the camper and go camping with Ralph and Luisa. It is good to get out and see different things, meet people, and see how people live.

Pete Abatielle called me from Bien Hoa, where he works, the other day. It was good to hear from him. He is thinking about extending and has a good job. He had a good time home and is anxious to get out of the Army and get back to teaching. I will be able to talk to him more, for he gave me his phone number.

Probably Sylvia already wrote and told you that Taffy got hit by a car. She is okay, but spent a few days at the animal hospital for a check up. The lady that hit her never even stopped.

Sylvia will be able to take a maternity leave for our baby. She will not lose tenure or position this way. She will have to go back to teaching for May and June, but we can work something out.

I am glad you told Mrs. Thomas. It has been a while since I heard from Marv [his best friend growing up].

The weather is changing a lot and we have a lot of sickness here. I am drinking a lot of Start, and so far I feel fine.

Take care, God bless.

Love,
Tom

September 18, 1971

Dear Mother and Dad,

I have received your letters and also my care package. Sylvia sent me peanut butter, but it broke. I am glad yours didn't. Now we can have a little party.

We really appreciate the book you sent, The Power of Positive Thinking. Steve is reading it now. I was looking through it last night. I am working on my correspondence course, so I told Steve he could read the book first. If you happen to come across any more of his books, I would like very much for you to send me the books. I like his philosophy, and I am trying to put it to every day living. It really works.

I represented Casualty Branch yesterday at the Soldier of the Month board. I applied positive thinking and took 2nd place. I was really pleased. I will get a check for $10, a three day R&R to China or Eagle Beach and represent the Adjutant General section at the division Soldier of the Month Board in October. Everything seems to be working out fine.

I hope you had a nice trip. I imagine you really enjoyed your stay with Donna and Charles.

I believe the Monsoon weather is here. It has rained for the last two days, with periods of clearing. It is hard to do your wash and get it dry.

Not much news over here. I don't believe I have written in a while. The days seem to go by pretty fast.

Take care, God bless.

Love,
Tom

October 1, 1971

Dear Mother and Dad,

I guess it has been a while since I wrote last. Not much has been happening and the time seems to be going by pretty fast.

I acquired a new job, besides working with letters of sympathy. I send out letters to men wounded in action. It is signed by the Commanding General and states that they receive the oldest medal, the Purple Heart, and he hopes they recover. It is just something to recognize that they were wounded for their country. It is a little extra work, but I don't mind.

I went before the next higher Soldier of the Month board today. I didn't win, but I don't mind. I really wasn't too excited about going before all these boards, but it didn't hurt me. It was an experience.

I will be going to China Beach for my R&R the 4th to 6th of October. It is my reward for placing second on the first board. It is in Danang, and I am hoping I can see Stevens. I won't have a lot of time, but maybe we can work something out.

The South Vietnam elections are 3 October. The military feels the NVA will start something, so the reaction force that I am on will have to sleep outside in a group, so we can be ready at a moment's warning. Phu Bai hasn't been hit in so long, that they can't remember when the last time was.

Dad, I was wondering if you wanted any more filters for your lenses. I could get some more daylight filters to protect the lenses from dust and scratches. They only cost a dollar or so. Let me know.

Take care and God be with you.

Love,
Tom

October 5, 1971

Dear Mother and Dad,

I am at the China Beach R&R Center, and boy am I enjoying myself.

I left Phu Bai at 10:30 AM and arrived in Danang at 11 AM. I flew by airplane, C-130. From the airport, I traveled by bus to get here.

It is just like the beaches in New Jersey. If it wasn't for the helicopters going over, it would seem like you were back home.

I have met three new buddies, and we do a lot together. I went swimming yesterday and in the evening we saw a floor show. They had a band and some dancers and a comedian. They were good. After the floor show, they had three movies.

The food is really good. You can eat all you want, but I think my stomach shrunk, and I can't eat that much in one sitting.

This morning we went swimming again and afterwards took a walk over to the PX. This afternoon we are just relaxing. There is another floor show tonight.

Tomorrow morning, I think I will try surfing.

I am taking a few pictures with my Instamatic. I will not need any more film. I still have this roll to finish up, and with the monsoons coming, there is little to take pictures of any way.

I have been working on my Army correspondence course this afternoon. It takes a while to finish it, so I have to keep at it.

Take care and God bless.

Love,
Tom

P.S. I got a letter from John the other day. It was good to hear from him. He is fine.

October 9, 1971

Dear Mother and Dad,

My R&R is over, but I really enjoyed myself. The last couple of days the weather was bad, so I did a little reading and shot some pool.

I was supposed to fly out on the 7th October, but the flight was canceled due to malfunction of the airplane, so I went back to the R&R Center. I got to see another floor show and a movie. I was manifested on a flight for October 8th, and I made it that time. I arrived in Phu Bai at about 3:30 PM, and it was cold and rainy. Last night, the wind really blew and I had to sleep under two blankets.

My work was piled up when I came in to the office today, so I had plenty to keep me busy. I got it all taken care of, so I am writing letters now.

Thank you for the care package. We had a little party last night, and I contributed to it. I also got a package from Sylvia and Donna.

I am glad you are able to go camping on the weekends. You seem to really enjoy yourself and that is only right, and I am glad.

Sylvia mentioned that she enjoyed your visit. I am glad that you got to see them and talk.

I got a letter from Coach McLaughlin today. There is still some trouble at the college, but he is out of it now and back at his old job.

I also got a letter from Donna. They are fine.

Is there anything in particular that you would like for Christmas?

Take care, and God bless.

Love,
Tom

P.S. Thanks for the Norman Vincent Peale book.

October 20, 1971

Dear Mother and Dad,

Not much happening in good old Phu Bai.

Thank you so much for the care package. We will be having a small Halloween party, and the goodies will come in handy.

I really appreciate the book you sent me How to Win Friends and Influence people. I started reading it today, and it is very good. There is so much I can learn from this book.

The positive thinking books are sure getting used. When one person finishes a book, it is passed on. We have five or six guys reading the books now.

We have been pretty busy lately, because of the increase in non-battle deaths. The office personnel is also getting low. We have six guys leaving within 20 days. We have only got 1 replacement. This means doubling up on jobs to get the work done. We should be getting replacements, but that takes time.

The one guy that we got is Charles Peach. He is from Pittsburgh. He is a good Christian fellow and will work out fine.

The night activities are somewhat limited. We usually go to the movies. Lately, I have been playing checkers and chess. It has been a long time since I played either a game, but I enjoy it.

The weather sure has been different lately. The sun only comes out for a few hours a day and then it rains most of the day.

We all had a flu shot last Monday. I am glad they gave me one.

Not much else going on.

Take care, and God bless.

Love,
Tom

October 24, 1971

Hi Mother and Dad,

We lasted through another typhoon. This one was the worst of all.

We had four sections of the roof from our billets blow off. My area kept dry, but some of the guys weren't as lucky. It rained for a good 24 hours.

Today, we only worked office CG's, because the electricity was off. The rest of us worked on the billets. We got ours fixed up again. It rained again today.

I understand that there is another typhoon on the way. I hope not, for some of them are pretty bad.

I got my next assignment: I will be going to Fort Carson, Colorado. I have mixed emotions over it. I am sure that the Lord will direct us.

Most of the guys that are leaving Vietnam in December are going to Fort Carson. It is possible that Pat is going there, too [and he did—they drove back to NY together]. I understand that Fort Carson is a liberal fort. They have used a lot of new ideas and are very progressive.

It doesn't look good for me coming home early. The drops are just not coming down. At any rate, I will be home for Christmas.

I got a letter from you today. I really do appreciate them. I got your suggestions for a Christmas gift. It sure is hard to get me much for Christmas. What I do need is civilian clothes, but I guess I should pick them out myself. I still am not sure what I should get Sylvia.

I sure am enjoying the book by Dale Carnegie. I am glad you sent it. It is going to help me in my career and everyday life.

I got a letter from Coach Houk the other day. He had to take a medical leave of absence from the college, because of his diabetes. He spent a short time in the Joslin Diabetic Clinic in Boston, and they were able to bring it under control. I feel sorry for Coach [Athletic Director at Bloomsburg University and also wrestling coach].

They are still having problems at the college, but I think they will get it cleared up.

We had our flu shots the other day. I am glad they gave it to us. The weather sure has been changing lately. Colds have been going around. So far, I haven't had one.

Take care and God bless!

Love,
Tom

November 2, 1971

Dear Mother and Dad,

Sorry I haven't written sooner, but time seems to go by pretty fast now.

I am extremely busy now. Our NCOIC (Non-commissioned Officer in Charge of Casualty Branch) went home to be with his wife, for she had a heart attack. Since I am the highest man on the E-5 list from Casualty Branch, I am now the acting NCOIC. It sure is a lot of work and responsibility, but I am managing. We have a new NCOIC coming in about a week. It is good experience.

We got two new clerks in the office, and both are from Pennsylvania. Paul Traiha is from Reading and Larry Peace is from Dubois. They are both good guys.

We have a new day room at Camp Campbell, and it is working out real well. We can play basketball, shoot pool, lift weights, or play table tennis. I have been playing pool lately, and I am getting better at it.

I really enjoyed the book by Dale Carnegie. Steve Pittman is reading it now. I plan to read it again, for it had so many good points that I would like to remember.

I haven't heard from John lately. He wanted me to order him some things. I hope he writes soon, for I am getting short on time.

No, drops haven't come down yet, so it looks like I will spend my whole time over here. At any rate, I will be home for Christmas.

Take care and God bless

Love,
Tom

November 13, 1971

Dear Mother and Dad,

Not much happening, but thought I would write.

We got four new guys in the office, so we are up to strength again. I got my replacement yesterday and have been teaching him my job. It won't be long until I will be home.

Bob Jacobs left about a week ago. He was a great guy. I hope I get a chance to see him again.

Steve Pittman, who works with me in Casualty Branch, will be going to Fort Carson, also. He DEROS's 3 January 1971. Some of the other guys that work with me are going to Fort Hood, Texas.

I got a letter from John. He didn't say much. He wanted me to order a few things for him. Since I was ordering him a watch, I decided to get one just like it. I needed another one, for the crystal on the one I have is cracked. The camera he wanted me to order for him has been discontinued. I hope I will be able to get him a camera when I pass through Japan.

Sylvia mentioned that our stereo equipment made it home in good shape. We will be able to enjoy it over the holiday season.

I am getting anxious to be home. It won't be long now. It sure is a nice time of the year to come home. We all have a lot to be thankful for!

I forgot to mention in my last letter that I went to the dentist. I didn't have any cavities and got my teeth cleaned. I was really happy.

We got a new NCOIC for Casualty Branch, so I don't have as much work to do.

I am presently reading a book about Jeanne Dixon. She is a remarkable woman.

I guess that is it for now. Take care, and God bless.

Love,
Tom

November 20, 1971

Dear Mother and Dad,

It was real good to get your letter and also a care package today. I also got a letter from the Dickinsons and Sylvia.

I don't believe how cold it gets here at night. The other night I went to bed with my pants on and had two blankets over me. The days are warm when the sun comes out again. This weather sure is mixed up.

I am working a new job. The hospital liaison for the 101st Division is on a 7-day leave, so I took his place. I am working at the 85th Evacuation Hospital, which is just down the road from Camp Campbell. My job is to keep track of all patients from the 101st. I enjoy the job, and I'm glad I got the opportunity to work with the men. I got a chance to meet a lot of different people.

Sylvia and I are letting God work things out for us concerning Fort Carson. She has mentioned going out with me, but we will have to discuss it and pray about it. Time will tell.

Thank you for the Reader's Digest. I was running out of reading material. I finished the book on Jeanne Dixon. It was quite good.

We went to the movies tonight, but it wasn't any good. I decided to come back to the billets and relax and read.

I guess that is all for now. Take care and God bless.

Love,
Tom

P.S. Enclosed you will find a copy of my orders.

November 25, 1971

Dear Mother and Dad,

Happy Thanksgiving!! Today has sure been a unique Thanksgiving. It is different from last year's.

The Commanding General of the 101st came to see the men at my hospital today. Everything went smoothly.

I had a real nice turkey dinner at the hospital, and then my ride came, so I am back at my billets earlier than usual. The guys at Casualty Branch are just running office CQ's all day, so they only have to work a few hours.

The weather is cold wet and rainy. It has been like this for about three days.

I guess the biggest news is that I got a 6-day drop. I will be leaving here Phu Bai around the 7th of December and leave Danang around the 11th of December. I should be in Trevorton around the 12th or 13th. I still have to process out at Fort Lewis when I get back to the states. Some of the other guys got bigger drops, because their regular DEROS were in January. Now, maybe I can do some shopping when I get home. I think Sylvia has most of the gifts bought.

I got two shots yesterday. I have to have my shot record up to date before I leave here. They were plague and cholera.

I have been nursing a bad cold, so while I got my shots, I also had a doctor check me out. He gave me some medicine, and I feel a little better.

We will probably go to the movies tonight for some excitement.

Take care, and God bless.

Love,
Tom

P.S. I suggest that you do not write after the 2nd or 3rd of December.

December 3, 1971

Dear Mother and Dad,

Today is my last working day. I have waited a year for this, and I will enjoy it. I will be leaving Phu Bai around 8 December to go to Danang. I will get a flight out of Danang to Fort Lewis on 11 December. I hope to be in Allentown airport on 12 December or 13th.

I can't imagine what happened to Pat. I have been trying to get a hold of the 1st Cav Division Casualty Branch, but so far I can't get them. These phones over here are really bad. Maybe Pat's parents have heard from him by now.

The weather is about the same over here, cold and rainy. We haven't had a nice day in about a week.

I have to go through a lot of processing before I get home, but it will be worth it.

This will probably be my last letter, for I will be home soon.

Take care!

Love,
Tom

Afterward

by Lynn Houston

My parents, Tom and Sylvia Houston, were married on May 23, 1970, and moved into an apartment they adored in Bloomsburg, Pennsylvania. It had parquet floors, built-in shelves in the dining room, and a fireplace in the bedroom. On Father's Day that June, my grandfather called and read them the letter that had come from the draft board. My mother remembers celebrating their first wedding anniversary without him: she went to dinner at the small restaurant area inside Kmart with her mom and the woman who babysat her as a child. With dad gone, mom spent holidays with her family and friends.

Although my mother missed my father, she was busy with her teaching job. Outside of those duties, she would also write him letters and send him homemade baked goods. My father said that they ate the baked goods quickly so that they had less to carry in their packs.

When my dad returned, Mom resigned from her teaching position in Pennsylvania. They packed up their blue Impala with baby supplies and drove to Fort Carson, stopping to see my father's older sister and her husband, Donna and Charles Moseley. Once they arrived in Colorado, they settled into an apartment off-post with beautiful mountain views. My mother stocked up on groceries and more baby supplies since I was due to be born mid-March.

When my father reported for in-processing, he found out that he was getting an early discharge. They gave him the option to re-enlist to allow his baby (me) to be born in an Army hospital. Instead, he took the early out. My mother remembers going around to the neighboring apartments and giving away most of the food she had just purchased.

My dad's childhood friend, Pat Manderville, had also been given an early out, so the three of them drove back home to NY, all squeezed across the front seat of the Impala since the back seat was filled with a crib and other baby items. They drove straight through, 27 hours. My dad then took a permanent substitute teaching position at Waverly High School where he taught in the business department from February to June.

I was born on a Tuesday afternoon in March, just as my dad got to the hospital after leaving school. The delivery cost them $1,500 in cash as neither of them had health insurance at the time. In May, my mom went to her mother's house in Trevorton, Pennsylvania, to take classes at Bloomsburg College for her master's degree. My father was going to join her after finishing the school year in June, but to do so, he had to ride his motorcycle through the flood of 1972, when the waters of the Susquehanna River crested at about 33 feet. He stopped to see a friend, Norris Rock, to get help figuring out which bridges and roads might still be open. A journey of three hours took him over five, but "come hell or high water" he was going to reach his wife and infant daughter!

That summer, my dad worked at the Bloomsburg swimming pool and completed the final classes for his master's degree. He applied for a teaching position at Marlboro High School in Marlboro, New York, after a friend told him about the opening. He interviewed and was offered the job on the spot. For the first few years of my life, we lived in a large apartment complex along 9W on the outskirts of Marlboro, until my parents borrowed $1000 for a down payment on a small house. Mom got help from a neighbor to watch me while she finished her master's thesis. My brother was born two years later.

My mom stayed home with us and brought in some extra income by offering childcare services for the neighborhood. You could do that in the '70s—no one required a business license or permit to operate an informal daycare out of your home. When we entered middle school, mom went back to teaching. She worked at St. Thomas of Canterbury Catholic Elementary School in Cornwall, New York, for seven years, then she took a job at the Newburgh Enlarged City School District, where she was an elementary school teacher for 20 years.

My dad worked at Marlboro High School for his entire career: he was a business teacher for 13 years and the assistant principal for 20 years. For many decades, he also ran a successful after school program teaching martial arts.

In retirement, they have both found fulfilling hobbies that help them feel close to their local community: my mother enjoys watercolor painting, and my father serves on the Board of Directors at Heritage Financial Credit Union.

My parents' 50th wedding anniversary fell on May 23rd, 2020, during the height of the COVID-19 pandemic. Interstate travel was difficult to impossible. No family members would be coming to town to attend the big party that had been planned. Even my brother, working as a pharmacist at the time in Pennsylvania, could not make it. Instead, it was just the three of us. We ate take-out from a local Italian restaurant. I read a poem. They told me stories about when they dated, and then they slow danced in the living room to their song, "The Impossible Dream." In the end, it was the perfect celebration.

Historical Context

Vietnamization (1969-1973)
One of the defining policies during the period of my father's letters was the strategy of Vietnamization, introduced by President Richard Nixon in 1969. Vietnamization aimed to reduce the U.S. military presence in Vietnam by shifting the burden of fighting to the South Vietnamese forces (ARVN). This process involved training and equipping the ARVN to take over combat duties, allowing the gradual withdrawal of U.S. troops.

By 1970-71, Vietnamization was well underway, with thousands of U.S. troops being withdrawn from Vietnam. However, as American forces were reduced, ARVN troops were expected to handle more combat operations, though they often lacked the same level of training, morale, and leadership. This shift may have been noticeable to my father as he saw fewer U.S. forces in the field and more South Vietnamese troops assuming combat roles.

Cambodian Incursion (1970)
In April 1970, President Nixon announced the invasion of Cambodia, a military operation aimed at cutting off North Vietnamese supply lines that ran through the Ho Chi Minh Trail, which passed through Cambodia. The incursion targeted communist sanctuaries in the country but also expanded the war into a neighboring country, leading to widespread protests in the U.S.

The Cambodian Incursion briefly increased the level of fighting in the region, particularly along the borders of Cambodia and South Vietnam, where my father was stationed. The expansion of the war was deeply controversial back home and fueled massive anti-war protests, including the tragic Kent State shootings in May 1970, where four students were killed by National Guardsmen during a protest.

The Laos Offensive and Lam Son 719 (1971)
In early 1971, the Laos Offensive, also known as Operation Lam Son 719, was launched by South Vietnamese forces with U.S. support to disrupt North Vietnamese supply lines in Laos. While U.S. troops were prohibited by Congress from entering Laos, American airpower and advisors supported the ARVN's efforts.

The operation ended in a significant failure for the South Vietnamese, as they suffered heavy casualties and were forced to retreat. This was a major blow to Vietnamization, as it exposed the weaknesses of the ARVN without direct U.S. combat involvement. During this period, U.S. forces like my father's may have been on high alert or directly involved in supporting operations related to the offensive.

Declining U.S. Troop Morale
By 1970-71, U.S. troop morale in Vietnam had declined significantly. Soldiers were increasingly aware that the war was unpopular at home, and the growing anti-war movement made many question their mission. Drug use, racial tensions, and incidents of "fragging" (the deliberate killing of unpopular officers) became more common.

My father's letters mention his unit having good discipline and camaraderie, but he also hints at the weariness of long deployments and monotony. This was a common sentiment among soldiers who, by this stage of the war, saw little progress in the conflict and were keenly aware that U.S. involvement was winding down.

Anti-War Movement and Protests in the U.S.
The anti-war movement in the U.S. reached its peak during the time my father was in Vietnam. Protests were held on college campuses, and public opinion increasingly turned against the war. The release of the Pentagon Papers in 1971 further fueled public disillusionment by revealing that the U.S. government had been less than forthcoming about the war's progress and prospects for success.

While my father doesn't discuss the protests directly in his letters, many soldiers were aware of the growing opposition to the war, and it influenced how they viewed their role and the broader mission in Vietnam.

Withdrawal and Transition (1971)
As U.S. forces were withdrawn, the remaining soldiers were often placed in defensive roles, protecting bases and supporting ARVN operations rather than engaging in large-scale offensive actions. The focus shifted to securing U.S. personnel and preparing for the eventual turnover of the war to the South Vietnamese.

My father's later letters reflect this transition, particularly his assignment to a rear position with clerical duties in the 101st Airborne Division. This was a common experience as more American combat units were drawn down, and those who remained were involved in less direct combat.

Notes

101st Airborne
The 101st Airborne Division, known as the "Screaming Eagles," was one of the most elite divisions in the U.S. Army during the Vietnam War. Originally a parachute division during World War II, the 101st later became an air assault unit, specializing in helicopter warfare and rapid deployment during the Vietnam conflict.

Airmobile
The 1st Cavalry Division became known as the "First Air Cavalry" after adopting airmobile tactics during the Vietnam War. These tactics allowed rapid troop deployment and support using helicopters, revolutionizing U.S. military strategy in jungle warfare.

Ashau Valley
The Ashau Valley was a critical battleground during the Vietnam War, located near the Laotian border in central Vietnam. The valley was part of the Ho Chi Minh Trail, a key supply route for North Vietnamese forces. Major battles, including Operation Apache Snow in 1969, took place in this area as U.S. forces attempted to cut off enemy supply lines.

Bien Hoa
Bien Hoa is located just outside Saigon (modern-day Ho Chi Minh City) and was the site of one of the largest U.S. airbases in Vietnam. It was a critical location for both air operations and troop deployment during the Vietnam War, frequently targeted by North Vietnamese forces.

Bob Hope
Bob Hope, the famous comedian, entertained U.S. troops overseas during numerous wars, including Vietnam. His shows,

known as USO (United Service Organizations) performances, were a morale booster for soldiers stationed far from home, often around holidays like Christmas.

Bonnie and Clyde
The 1967 film *Bonnie and Clyde* was a major cultural touchstone, depicting the infamous outlaw couple during the Great Depression. Its blend of romance, violence, and rebellion against authority captured the spirit of the 1960s, influencing the counterculture and breaking new ground in American cinema with its gritty realism.

Buddhist Temples
Buddhist temples in Vietnam often feature traditional architectural styles with curved roofs, stone statues of bodhisattvas, and intricate carvings. Temples serve not only as religious sites but also as cultural landmarks, playing a key role in the spiritual life of many Vietnamese people. These temples frequently incorporate symbolic elements such as lotus flowers, representing purity, and guardian statues like lions or dragons.

Cam Ranh Bay
Cam Ranh Bay is a deep-water port located on the southeastern coast of Vietnam. It was one of the most important logistical hubs for U.S. forces during the Vietnam War, serving as a major base for military supplies and troop arrivals. The bay's strategic location made it ideal for supporting combat operations in the region.

Central Highlands
The Central Highlands of Vietnam were a key strategic area during the Vietnam War. This rugged, mountainous region was heavily contested by both U.S. and North Vietnamese forces. The area was strategically important due to its proximity to the Ho Chi Minh Trail and its elevation, which allowed control over

key transportation routes. It was the site of numerous battles between U.S. and North Vietnamese forces. This forested region of Vietnam is known for its dramatic landscape and terraced fields.

Claymore Mine

The M18 Claymore mine is an anti-personnel mine used by the U.S. military, first deployed during the Vietnam War. It is designed to project shrapnel in a fan-shaped pattern when triggered, often used for defensive purposes in ambushes or to secure perimeters. It became a staple of Vietnam-era combat and is instantly recognizable by the phrase "Front Toward Enemy" engraved on the front.

Combat Infantry Badge (CIB)

The Combat Infantry Badge (CIB) is awarded to U.S. Army infantrymen who have engaged in active ground combat. During the Vietnam War, this was a prestigious recognition, symbolizing the dangerous and direct combat situations soldiers faced while serving in the conflict.

Combat Rations (C Rations)

Combat rations, or "C rations," were pre-packaged meals provided to soldiers during the Vietnam War. These rations typically included canned meats, fruits, and desserts, as well as crackers, cigarettes, and a small accessory pack with matches and gum. They were replaced later in military history with the more modern MREs (Meals Ready to Eat).

DEROS Center

DEROS stands for "Date Eligible for Return from Overseas." The DEROS center was where soldiers in Vietnam waited for transportation home once their tour of duty was complete. It was a major milestone for soldiers who had finished their service in the war zone and were preparing to return to the United States.

Early American Style
The "early American" or "colonial" style was a popular home decor trend in mid-20th-century America, evoking the aesthetics of the early United States. It often featured symbolic items such as eagles (a patriotic emblem) and utilitarian farm tools like oxen yokes, reflecting a simpler, pastoral time. Colonial-style homes were typically painted in muted, natural tones, and this decorative approach became a symbol of American heritage during the post-World War II period.

Fathers and Sons by Ivan Turgenev
Fathers and Sons is a classic Russian novel by Ivan Turgenev, first published in 1862. The novel explores generational conflict, social change, and the challenges of youth. Reading literature like this was common among soldiers looking for distraction and intellectual engagement during downtime in Vietnam.

Fort Dix
Fort Dix, located in New Jersey, was a major training center for soldiers during both World War II and the Vietnam War. It gained a reputation for its rigorous basic training programs, and units like Charlie Company often became known for their high standards and discipline.

Graves
In Vietnamese culture, ancestral graves are often elaborately decorated, reflecting the importance of ancestor worship. The dragon, a symbol frequently seen in Vietnamese funerary art, represents strength, prosperity, and protection. Such graves are common in rural Vietnam, and they feature traditional architectural elements like ornate pillars and stone walls.

Huế
The city of Huế in central Vietnam is renowned for its historic

and cultural significance, particularly as the imperial capital of the Nguyễn Dynasty. It is home to many Buddhist statues, temples, and pagodas, including the famous Thien Mu Pagoda. Huế also saw significant fighting during the Vietnam War, especially during the 1968 Tet Offensive.

Log Day
"Log day" was the day soldiers in Vietnam received supplies, mail, and sometimes hot meals after spending days in the field. The term "log" comes from logistics, as this was the day when logistical support was delivered to troops. For soldiers, it was often the only chance to get clean clothes, new supplies, and letters from home, making it feel like a small holiday.

Long Range Patrol food (LRP's)
Long Range Patrol rations, or LRPs, were dehydrated meals developed for soldiers on extended missions in Vietnam. Unlike the more common C rations, LRPs were lightweight and required the addition of hot water to rehydrate. They were specifically designed for troops who needed to travel light while spending long periods in the field.

M-60 Machine Gun
The M-60 is a belt-fed, gas-operated machine gun that was standard issue for U.S. forces during the Vietnam War. Known for its firepower and reliability, it earned the nickname "The Pig" due to its heavy weight (approximately 23 pounds) and bulk. It was commonly used by infantry units and played a crucial role in both offensive and defensive operations.

Monopoly Money: Military Payment Certificates (MPC)
The currency referred to in the letters is Military Payment Certificates (MPC), used by U.S. soldiers overseas during the Vietnam War to avoid black market activity and control inflation. The colorful, fake-looking currency was often compared to the play money from the popular board game

Monopoly, which had been a household favorite in America since its release in the 1930s.

Movies
During the Vietnam War era, movies reflecting the social and political turmoil of the time, such as *The Graduate* (1967) and *Easy Rider* (1969), resonated with audiences, especially younger generations. Many films from this period captured themes of rebellion, counterculture, and opposition to the war, mirroring the campus demonstrations that were occurring across the U.S.

Action films featuring heroic figures and intense battle scenes were also common forms of entertainment for soldiers. Films like *The Green Berets* (1968), starring John Wayne, were particularly popular, as they depicted the war in a positive, patriotic light, aligning with the U.S. military's values at the time.

Phu Bai
Phu Bai was a major U.S. military base located near the city of Huế in central Vietnam. It served as a staging point for operations in the northern part of the country, particularly during the Tet Offensive in 1968. Phu Bai was also known for housing the 101st Airborne Division, one of the key units during the war.

PX
A "PX" (Post Exchange) is a retail store located on military bases where service members can purchase goods at reduced prices, similar to a civilian department store. Soldiers could also order items through the PX catalog and have them shipped to their location. This service was a valuable way for soldiers stationed overseas to acquire personal items, gifts, or necessities.

Reader's Digest
Reader's Digest was one of the most popular magazines in America during the 20th century. Known for its condensed versions of articles and books, it was widely read by both civilians and military personnel. It often featured stories of American values, family, and perseverance, resonating with soldiers during times of war.

Rest and Recuperation (R&R)
R&R, or "Rest and Recuperation," was a break granted to soldiers serving in combat zones, allowing them to spend time in relatively peaceful areas or even in other countries. Common R&R destinations for soldiers in Vietnam included places like Thailand, Hawaii, and Australia, where they could relax away from the war for a few days before returning to duty.

Rotary Club
The Rotary Club is a global service organization made up of business and professional leaders. Many members pay regular dues or make contributions to support charitable initiatives.

Saigon
Saigon, the capital of South Vietnam during the Vietnam War, was a central hub for U.S. and allied forces. It served as the administrative and military center of the South Vietnamese government and was the scene of significant combat during the war, especially during the 1968 Tet Offensive.

Song Bé
Song Bé Province, located near the Cambodian border, was a contested area during the Vietnam War. It was a rural region with dense jungles and rubber plantations, often used as staging grounds by North Vietnamese forces infiltrating from

Cambodia. The proximity to the border made it a site of frequent skirmishes and patrols by U.S. forces.

Stars and Stripes
Stars and Stripes is the military's independent newspaper, providing news to U.S. service members stationed around the world. It covered both global events and stories from the front lines, helping soldiers stay connected to events back home and developments in the war. It was widely read by troops during the Vietnam War.

Sundy Punch
"Sundy Punch" was likely a small fire support base or temporary military position during the Vietnam War. Fire support bases like Sundy Punch were established by the U.S. military to provide artillery support and protection for infantry operations. Many of these bases were given informal or temporary names and were often dismantled or handed over to South Vietnamese forces as U.S. involvement in the war began to wind down.

Troop Carrier
During the Vietnam War, helicopters such as the UH-1 Huey were widely used as troop carriers. These helicopters revolutionized military operations, allowing soldiers to be rapidly deployed and extracted from difficult terrain. The ability to use helicopters for transportation and medical evacuation was a defining feature of the Vietnam conflict, leading to the term "airmobile" units like the 1st Cavalry Division.

www.ingramcontent.com/pod-product-compliance
Lightning Source LLC
Chambersburg PA
CBHW070140080526
44586CB00015B/1774